WHEN MUSCLE PAIN WON'T GO AWAY

WHEN MUSCLE PAIN WON'T GO AWAY

3rd Edition

The Relief Handbook for Fibromyalgia and Chronic Muscle Pain

by

Gayle Backstrom

with

Dr. Bernard Rubin

Professor of Medicine
University of North Texas
Health Science Center of Texas
at Fort Worth

TAYLOR PUBLISHING COMPANY
Dallas, Texas

To
Judith Rosen and Carol Yager
and to all of the
Support Group Leaders
who give so much of themselves
in spite of their own fibromyalgia
so that others may learn more about FM

Copyright 1992, 1995, 1998 © Gayle Backstrom
All rights reserved.

No part of this book may be reproduced in any form or by any means without written permission from the publisher.

Published by Taylor Publishing Company
1550 West Mockingbird Lane
Dallas, Texas 75235

Illustrations by David Hunter unless otherwise noted.

Library of Congress Cataloging-in-Publication Data

Backstrom, Gayle.
    When muscle pain won't go away: the relief handbook for fibromyalgia and chronic muscle pain / by Gayle Backstrom with Bernard Rubin.—3rd ed.
        p.   cm.
    Includes bibliographical references.
    ISBN 0-87833-998-1
    1. Fibromyalgia. 2. Myalgia. I. Rubin, Bernard R. II. Title.
RC927.3.B33   1998
616.7'4—dc21   98–7860
                                                          CIP

Printed in the United States of America
10 9 8 7 6 5 4 3 2 1

Acknowledgments

I would like to thank Judith Rosen and Carol Yager of Dallas, Support Group Leaders for the Dallas Fibromyalgia Support Group for all of their help in sharing their notes from several conferences that they attended and I could not, plus for all of their kind words and thoughts. I would also like to thank my agent, Jim Donovan. He believed in my book enough to buy it when he was an editor at Taylor Publishing and later, believed enough in me to take me on as a client when he formed his own literary agency. Jason Rath has worked with me on this third edition and I want to thank him for all of his help. Also Camille Cline, my current editor who stepped in to see the book to print. I also appreciate the great work by Carol Trammel, the art director at Taylor, for making the book look reader friendly; and thanks also to Michelle Justiss.

I would like to thank Brenda Napier and other members of the Fibromyalgia Association of Greater Washington; Rae Marie Gleason of the National Fibromyalgia Research Association of Salem, Oregon; and Mary Anne Saathoff, RN, BSN, of the Fibromyalgia Alliance of America, Columbus, Ohio, for their help in sharing information and for all of the hard work they and other members of their associations do, not only in helping increase awareness of fibromyalgia but also for their efforts in funding research.

I want to thank my sister, Phyllis Knox, who helped in typing for this edition; David Hunter for his illustrations; and Justin Harden, who not only is my massage therapist, but who also kindly helped with some typing of the Positive Tips. I would like to thank everyone who sent me their Positive Tips for Living with Fibromyalgia (noted in shaded boxes throughout book). I have always tried to think in positive terms and these individuals willingly shared their own tips with me and all readers of this book.

Again, it has been friends who supported and encouraged me through the research and revisions, but this time, I also had my sister, Phyllis, and her daughter and son-in-law, Tammie and Ben Reyes with their two daughters, Ashley and Brianna, who have been there for me. (Ashley thinks I never eat or sleep or do anything except sit at the computer.) Thanks also to friends Donna Bell, Joy Wright, Linda Erickson, Billie Cantwell and, of course, Maurine Burnett, along with a couple of new friends, Teri Burgoon and Sharon Humpert, who carried my load at the Denton Citizens Police Academy Alumni Association while I worked on this.

Contents

Introduction

As so many people do when trying to set a time frame to their life, I mark time as before and after, specifically, before my major flare-up in 1986-87 and after. That is only natural when a particular event has had such a tremendous impact upon a life. In 1987 after a five-week hospital stay and then subsequently having to give up my full time work, I faced my future with a lot of uncertainty and fear. I was experiencing almost constant high levels of pain and exhaustion which kept me from not only pursuing a career but also any leisure activities that demanded much energy—make that even mild energy. I felt that I stood on the brink of a deep chasm with what I considered a "normal life" on the other side and no bridge in sight.

By the time I wrote the first edition of *When Muscle Pain Won't Go Away*, I had begun to build my own bridge. It was slow work but not lonely work. Dr. Bernard Rubin contributed a great deal of solid timber to help form the basic structure. That "timber" was not in the form of a cure for my fibromyalgia, there was no cure then, and there still isn't one today, YET. But what Dr. Rubin gave me was the gift of a caring, knowledgeable doctor who believed in me and who gave me the beginnings of understanding about fibromyalgia.

He encouraged me, not only as a patient to educate myself about FM but also as a writer to pass the knowledge that I gained on to others. By nature, I am a nosy person, always trying to learn more about a wide range of subjects. In writing this book, I had a "legitimate" reason to dig deep into all of the areas that I was facing with a chronic illness as puzzling and frustrating as fibromyalgia. So I gathered knowledge and used it to form the major supports of my new bridge. Because I not only had to research the subject, but also digest it as much as possible and then turn around and explain it to an audience of others who also had FM, I believe that I gained even more understanding than I might have if it were only for me.

I have had quite a number of others helping me in building the bridge which is in actuality, my learning how to live with FM. There have been psychologists, some that I saw long before I realized the full impact of FM, others I saw along the way. There were physical therapists, occupational thera-pists, nurses, and massage therapists who helped me physically and mental-ly. There were friends who have stood by me, one who saved my life, and oth-ers who have taught me that it is okay to accept help when I needed it.

There are close spiritual friends whose faith has shored up mine when I faltered and doubted in my ability to accept the changes in my life. There are both volunteers and police officers from the Denton Police Department who accept that I have something to give despite the limitations of time and energy.

Perhaps one of the biggest surprises was the responses I have received because of this book. Each letter thanking me for writing the book and telling me how it has helped has provided more links to the walkway of my bridge, giving me solid support to walk on.

Now just over eleven years after I left my last full time job and as I write this third edition to the book, I realize that my bridge is still being built. Oh, I have made a great deal of progress on it and at first glance, it seems to be a solid structure. In other words, I often feel as if I know all I need to about liv-ing with FM, at least for me, personally. And yet, each day, I learn something new, not just the medical information which I researched to go into this edi-tion but the practical, day-to-day aspects of life.

Basically, what I have learned is that my bridge will never be complete and the reason is not that I can't live with FM. The reason is that life is a jour-ney, not a destination. Until I die, I will continue to be learning new aspects, aspects involved with the day-to-day experience of living with a chronic ill-ness. But that is okay. If I ever thought I knew everything, it would be evident that in reality, I knew nothing. Despite the uncertainty of the FM itself, I am certain about one thing. I know that I can live with it, as long as I remain flex-ible and open and am willing to grow with the experience. And as for that "normal life" I saw across the chasm of pain, fatigue, and frustration? Well, I've learned that "normal" is what you make it and besides: Who wants to be normal anyway?

Gayle Backstrom, Denton, Texas

ONE

What is Fibromyalgia and Why Do I Hurt All the Time?

"I hurt all over, just like the flu but worse."

"I'm so tired that I feel as if I could just melt into the mattress and never have the energy to move."

"Everyone tells me I'm looking great so I must be feeling okay. The fact is that I'm in a lot of pain. Why can't they see that?"

"I've seen fifteen doctors and they've run every kind of test imaginable but nothing has shown up. Everything looks normal. I must be going crazy. Actually, that's what several doctors have told me. It's all in my head. There's nothing wrong with me. If that's true, then why do I feel so bad?"

The widespread musculoskeletal pain, fatigue, and poor sleep of fibromyalgia can be traced back to ancient times. David could have been describing it several times in Psalms. According to Dr. Xiao-Ming Tian, an expert on acupuncture and Chinese medicine who spoke at the Mid-Atlantic Conference on Fibromyalgia Treatment held in 1996, it was even discussed in ancient Chinese writings. It has been known by many names and although today's term "fibromyalgia syndrome" or (FM) is the best yet, it still doesn't give the whole picture of this puzzling, painful and chronic condition. (*Fibro* refers to the tissues, tendons, and ligaments; *my* to muscles; and *algia* means pain.)

It is often hard to explain to others just how bad we feel, because outwardly we often don't look ill at all. Sometimes we would like to have the outside show just how bad we really do feel as the mirror shows.

Traditionally, Western physicians have divided medicine into two areas. In illnesses, there is either a physical problem—infection, injury—to the body, or a psychological problem. If no evidence of abnormality shows up on lab tests or x-rays, then the problem must

My wife has FM. I (we) have learned that life goes on. Often, we make plans to do things together. Often, because of her condition she can't follow through. However, she encourages me to go ahead and not feel guilty, nor does she lay guilt. I (we) don't like it, but it's how things are. —Bob and Donna

be psychological. Therefore, according to traditional medicine, when a person comes to a doctor complaining of fatigue and widespread muscle pain, along with a number of other vague problems, and nothing can be found on the usual tests, the person is obviously suffering from a psychological disorder. In fact, many of the symptoms of fibromyalgia mimic those of depression or other mental illnesses and this complicates the acceptance of a physical condition. There have been thousands of patients who have been told that their problem is psychological and that they must "learn to live with it" or "just forget it."

Even when a doctor believes his patient, it is hard to pin down the cause of these vague symptoms. Fatigue and pain are very much subjective symptoms. Many caring doctors have conducted expensive, invasive tests to no avail, or unsuccessful surgeries have been performed only to have the individual continue to experience chronic pain—sometimes worsened both by the stress of the surgery and the failure to get better.

How is a Diagnosis Made?

When looking at any health problem there are certain questions asked. What are the symptoms? What causes it? What factors have an impact on it—what makes it worse, what makes it better? Who gets it? What is the prognosis or long term outlook? Is there a cure? What kind of treatment will work?

By the mid-1970s several doctors began to realize there might be more going on in patients with chronic fatigue and pain than just malingering. Dr. Hugh Smythe and Dr. Harvey Moldofsky at Wellesley Hospital in Toronto, Canada, conducted a number of studies on sleep patterns and fibromyalgia beginning in the mid-1970s and into the 1980s. Dr. Moldofsky was able to reproduce the muscle pain of fibromyalgia in normal, healthy individuals by disrupting the stage-four or non-rapid eye movement restorative sleep, except for one very aerobically fit person. Once allowed to sleep undisturbed, the muscle pain disappeared.

These early studies also showed the presence of *tender points*, which have become one of the keys to diagnosis of fibromyalgia. These tender points are specific locations on the body that are very painful upon palpation or pressure. In the normal healthy subjects of the sleep studies, the tender points disappeared along with the muscle pain and fatigue once restorative sleep was achieved. One of the questions that still remains about fibromyalgia and its symptoms is which comes first—the pain, the tender points, or the disrupted sleep?

The first step in treating any health problem is to arrive at a correct diagnosis and devise a successful mode of treatment. Individuals with fibromyalgia find it difficult enough to live with this confusing, painful, and frustrating condition without having to defend the reality of the diagnosis. Unfortunately, not everyone is able to obtain a correct diagnosis without some hard searching for a physician who understands how to diagnose FM and who can then work with the patient in developing a treatment program.

As fibromyalgia is becoming more widely known (articles are appearing in medical journals directed towards family practitioners, nurses, and other health care providers), the odds are better now that an individual will get a correct diagnosis. There are still a number of physicians, however, who insist that FM is a psychogenic disorder.

An Australian physician accused Dr. Robert Bennett, one of the leading experts on fibromyalgia, and other "devotees of fibromyalgia" of being "cocooned" within the confines of the *Journal of Musculoskeletal Pain,* and urged them to rejoin the scientific community. Dr. John Quintner stated that otherwise "fibromyalgia will continue its current downward spiral to becoming a derogatory and discredited diagnosis, akin to 'neurasthenia' and 'hysteria' of the late nineteenth century."

Dr. Quintner's letter and Dr. Bennett's response were printed in the *Journal of Musculoskeletal Pain* (Vol. 5, No. 3, 1997). I loved Dr. Bennett's response. Before referring to 15 papers on abnormalities found in individuals with fibromyalgia and recommending Dr. Quintner study them, he asked if perhaps the findings of Drs. Vaeroy and Russell (editor of the *Journal of Musculoskeletal Pain*) regarding elevated levels of substance P in the cerebrospinal fluid of fibromyalgia patients were a psychic phenomenon. Russell and Vaeroy's work have been verified by other researchers.

Interest and belief in fibromyalgia has grown over the last two decades. In 1990, the American College of Rheumatology (ACR) finally established the diagnostic criteria for fibromyalgia syndrome. (A syndrome is a condition that has a specific set of symptoms, yet no known specific disease to cause them.) To meet the criteria for a diagnosis of fibromyalgia, an individual has widespread, diffuse pain occuring on both sides of the body as well as the upper and lower parts. In other words, there must be pain on the left and right side of the body, and it must have occurred both above and below the waist. The pain must persist at least three months.

Although various researchers found hundreds of tender points around the body, the ACR decided that 18 were important. When these points are pressed with the finger or with an instrument called a dolorimeter, at least 11 of them must be painful. Most individuals with fibromyalgia are unaware of these tender points until they are pressed by the examiner.

The American College of Rheumatology's Criteria for Classifying Fibromyalgia

1. History of widespread pain

    Pain is considered widespread when all of the following are present: Pain in the left side of the body, pain in the right side of the body, pain above the waist, and pain below the waist. In addition, *axial skeletal* pain—(pain in the *cervical* spine (neck), chest, *thoracic* spine (upper back), or lower back)—must be present. In this definition, shoulder pain and buttock pain qualify as pain for each involved side. Lower back pain is considered lower segment pain (below the waist).

2. Pain in 11 of 18 tender point sites upon *digital palpation* (pressure applied with the fingers).

    Pain, on digital palpation performed with an approximate force of 4 kg, must be present in at least 11 of the 18 designated tender point sites. For a tender point to be considered positive, the subject must feel pain, not tenderness, upon palpation.

    For classification purposes, fibromyalgia is diagnosed if both criteria are satisfied. Widespread pain must have been present for at least three months. The presence of a second clinical disorder does not change the diagnosis of fibromyalgia.

Courtesy of Dr. Frederick Wolfe; reprinted with permission from *Arthritis and Rheumatism*

Prior to the adoption of these criteria, fibromyalgia was considered to be primary, secondary, or concomitant. *Primary FM* applies when fibromyalgia was the only rheumatic condition the patient had. *Secondary* or *concomitant FM* indicated the individual might also have rheumatoid arthritis, lupus, osteoarthritis, or some other form of arthritis. Those researchers who worked to develop the final criteria believed that it was not important if the individual had another condition; both the fibromyalgia and the other condition must be treated. Quite often individuals who have FM with another rheumatic condition find their arthritis symptoms are worse because of the FM, so it is important to treat both.

Even though the adoption of these criteria for diagnosis was important, many fibromyalgia researchers believe they need to be fine-tuned. Perhaps it is inevitable that this should be the case in a condition as puzzling as fibromyalgia. The current criteria, however, are sufficient for most doctors to reach a diagnosis of FM without large numbers of expensive and invasive tests.

What are the Symptoms?

Some of the other symptoms found in fibromyalgia include stiffness, headaches, numbness or tingling in the hands or feet, chest pain, dizziness, problems with balance and cognitive dysfunction such as short term memory loss, difficulty in finding a particular word, or difficulty in concentrating. A number of conditions are often found in people with FM, including depression, irritable bowel syndrome, other digestive disorders, irritable bladder, painful menstrual periods, and sensitivities to various stimuli from drugs to noise. *Dysfunctional syndromes* are those where there is no proven organic cause; this definition applies to fibromyalgia as well as to several of these other conditions.

It is not unusual to find fibromyalgia patients who have one or more rheumatic conditions like rheumatoid arthritis, lupus (systemic lupus erythematosus), or osteoarthritis.

Although those who study fibromyalgia agree that it is not solely a psychological illness, most agree that there is a group of individuals with fibromyalgia who also experience anxiety or depression. This percentage of individuals usually averages about 30 percent, a figure comparable with individuals in the general population. Whether the depression comes from living with a chronic pain condition, which can have a major impact on the person's quality of life, or from a separate cause, most researchers agree that this depression should be treated as well.

In the last ten years, research on fibromyalgia has grown at a tremendous rate, yet more questions have been raised than answered. What has become evident is that those with FM are a diverse lot. Some individuals have symptoms that have only recently developed and who receive relief from quick treatment. Others have had symptoms for up to 30 years or more; within that group, some have very light symptoms while others are unable to carry out normal daily activities because of severe pain and fatigue.

Some individuals can point to a specific incident—an auto accident, viral infection, surgery, even a severe emotional strain—which precipitated the advent of symptoms. In these people, there was either an emotional or physical stressor of some sort. (Many people don't realize that even an illness such as flu can be a stressor for the body.) For other FM patients, symptoms just slowly crept up on them until one day they realized it had been years since they had energy or a day without pain.

Many researchers believe that fibromyalgia patients can be broken into subgroups and that until those subgroups are recognized, studies will continue to come up with conflicting results. These subgroups could explain why everyone doesn't get the same set of symptoms or the same response to treatment.

Modulating Factors for Fibromyalgia

Aggravating Factors	Relieving Factors
Cold or humid weather	Warm, dry weather
Nonrestorative Sleep	Hot showers or baths
Physical/mental fatigue	Restful sleep
Excess physical activity	Moderate activity:
Physical inactivity	stretching exercises and
Anxiety/stress	massage

Courtesy of Dr. P. Kahler Hench. Reprinted with permission from *Rheumatic Disease Clinics of North America.*

Modulating Factors

The list of symptoms involved with FM is long and varied but not everyone has or will have every one of them. Certain conditions do have an impact on the severity of symptoms, making them sometimes worse and sometimes better. The symptoms of FM wax and wane, just as the symptoms of many other illnesses do. Some days are very bad and some days are fairly good, often with no apparent reason.

Exacerbating factors include the weather; cold and conversely heat; inactivity; too much activity; stress, whether physical or emotional; nonrestorative sleep; and physical or mental fatigue. Cold, damp weather has been accused of making an individual feel worse but it appears that it is more a case of changes in the weather. The change in barometric pressure when a front moves in can signal increased pain and achiness by affecting the fluid within the body's cells. Sitting in a draft from an air conditioner can also increase the pain of FM, and many people find they cannot sleep with an air conditioner blowing directly into their bedroom. Very hot weather can also leave some individuals feeling ill, although the reason why is not certain.

Although many can understand that over-activity could cause pain, it's harder to accept that inactivity can also cause FM symptoms to worsen. However, an individual's level of aerobic fitness does have an impact on symptoms. At one time it was thought that individuals with fibromyalgia were simply "deconditioned" or had become "couch potatoes." While that theory is no longer popular, it is known that when an individual begins very mild

First you must learn to help yourself to cope, then you will be able to laugh at things and relate to others. Loving is giving funny sayings to others. —Jessie

activity, the body produces natural endorphins (the body's natural painkillers) and they tend to feel better. Exercise requires care and balance, because strenuous exercise after a long period of inactivity or deconditioning may lead to muscle injuries.

Physical and mental fatigue can keep the FM patient from sleeping well. Because nonrestorative sleep is already a factor in FM, additional fatigue helps to perpetuate a cycle of poor sleep, more pain, and more sleepless-ness. Although nonrestorative sleep is known as a factor in FM, what isn't known is which came first: the pain or the poor sleep. While one of the major efforts of treatment is to improve the quality of sleep, not everyone is able to obtain those benefits; they become caught up in an endless cycle. Beyond the intrusion of pain into the stage-four sleep, other factors in the bedroom may impact the quality of a night's sleep. These can include the mattress and pillows on the bed, the temperature of the room itself, any noise which disturbs the sleeper, and the presence of animals in the room or on the bed.

> *I love my cervical pillow. It helps the neck pain and it's worth the cost to get a good one! I make copies of my FM information and give it to my primary care physician. —Jeanne*

Stress also has a strong impact on FM and its symptoms. This is not surprising because we know that stress affects everything from blood pressure to gastrointestinal problems. However, just as negative stress can have an impact, so can positive stress. Sometimes our bodies don't know the difference between positive and negative excitations, giving the same physical response for both. Examples of positive stress include the following: a promotion with increased responsibilities, getting married, a move to a new neighborhood, the addition of a child to the family whether by birth or adoption or even retirement. The recognition that stress can make FM symptoms worse does not mean that it is psychological. There is more and more evidence that the link between mind and body is much stronger than we thought. What affects the mind affects the body. Negative stressors may include anything from financial difficulties to an unhappy marriage, from a troublesome job to a high neighborhood crime rate.

If certain factors make FM symptoms worse, is there anything that can make them better? Yes. Those with FM typically find that their symptoms are eased by warm, dry weather, hot showers or baths (or a session in a hot tub), restful sleep, and moderate activity including stretching exercises and massages.

Who Gets FM?

Fibromyalgia is found in almost every country and in every culture. No one knows exactly how many people have fibromyalgia although it is one of the most common conditions seen in rheumatology clinics and is considered the most common cause of chronic pain. Early studies looked at the incidence of fibromyalgia patients seen in the clinical setting and found that between 2 and 6 percent of all patients seen in primary health care had fibromyalgia; the figure rose to about 20 percent of those in rheumatology clinics. Studies conducted in the general population estimate fibromyalgia cases to be between 2 and 5 percent. Conservatively, around 10 million Americans have fibromyalgia.

Women make up the majority of FM patients, around 85 to 90 percent, but researchers can't pinpoint the reason. Fibromyalgia has been found in children, but the most common age range is from 30 to 80. Fibromyalgia is more common in older individuals. A study conducted in Wichita, Kansas found that approximately 7 percent of women over 70 have FM.

Many adults remember having "growing pains" when they were young without having a clear diagnosis. There is growing evidence that there is a genetic predisposition to getting FM; it runs in families and often relatives of individuals with FM suffer from depression or alcoholism. Research linking fibromyalgia to neurotransmitters within the central nervous system may be the answer to why these conditions seem to be related.

What Can I Expect?

This is a chronic condition. As yet, there is no cure and only a few persons have had complete remission of all symptoms. There is no evidence, however, that fibromyalgia develops into another more serious illness. Perhaps the most important fact to remember, once you obtain and accept the diagnosis of FM, is that you must learn how it affects your body.

Everyone is an individual, and everyone has a different level of involvement. Do not compare yourself with others; there will always be some who will be better and some who will be worse than you. Learn to listen to your own body to determine your limits. Some people will have only minor adjustments to make; others require major lifestyle changes. Some will become disabled. Dr. Frederick Wolfe, at the University of Kansas School of Medicine in Wichita, Kansas, reported on the findings of a survey of self-reported work disability in *Journal of Rheumatology*, June 1996. The study begun in 1988 covered 1,604 patients seen in six rheumatology centers and showed that overall 26.5 percent of the patients in the study were receiving at least one form of disability payments when Social Security Disability and other sources of disability payments were considered. The figures from the six different centers varied quite a bit regarding disability awards and self-reported disability, and the variations could have been because of clinic referral patterns, physician beliefs, or socioeconomic status. Whereas earlier figures in Wichita had less than 25 percent receiving Social Security Disability that figure rose to 46.4 percent. Readers must remember that those individuals seen in a specialty clinic such as a rheumatology clinic will represent those who have more severe symptoms.

Looking at it a different way, 64 percent were able to work all or most of the time, and 70 percent were employed or were homemakers. Dr. Wolfe and the other researchers recommended further study to determine how differences among the six centers could affect the true rate of disability.

There are many FM individuals in the general public who do not see any healthcare provider for their symptoms. Not everyone will become disabled, nor will everyone find it impossible to continue the activities of daily living. But for those whose lives have had to take a new direction, we will address these issues later in great detail. Just don't assume that a diagnosis of fibromyalgia is the end of your world as you know it.

Several instruments have been developed to help determine just how much FM has impacted the quality of your life. Fibromyalgia has been shown to have a more negative impact on quality of life than other chronic health conditions. While no one is quite sure why this is the case, questionnaires are used to help not only in the initial assessment of impact but also in any progress following treatment.

Many individuals do report *functional disability*, the inability to continue with their job or their daily activities. Estimates vary on how many people are disabled and not everyone agrees on how to approach the issue. A majority of individuals can continue their activities with some minor lifestyle adjustments, but there are some who can no longer support themselves. While some doctors advise individuals not to apply for disability because of the very negative aspects of the experience, others recognize that some people simply cannot continue to hold a job. You should try as hard as possible to

Doctor's Questionnaire for Determining the Severity of Your FM and Charting Your Progress

Were you able to:	Always	Most Times	Occasionally	Never
Do shopping	0	1	2	3
Do laundry with a washer and dryer	0	1	2	3
Prepare meals	0	1	2	3
Wash dishes/cooking utensils by hand	0	1	2	3
Vacuum a rug	0	1	2	3
Make beds	0	1	2	3
Walk several blocks	0	1	2	3
Visit friends/relatives	0	1	2	3
Do yard work	0	1	2	3
Drive a car	0	1	2	3

Courtesy of Dr. Robert Bennet, Carol S. Burckhardt, R.N., Ph. D, and Sharon R. Clark, R.N. Reprinted with permission from the *Journal of Rheumatology*.

continue working, not only for the financial returns (after all, this world operates on a money system, whether we like it or not), but also for all the other reasons that we work—our sense of self-worth and social contact with other people—among others.

Studies show that over an extended period of time, the symptoms of fibromyalgia remain fairly stable. There have been some cases which have gone into remission, which can last from months to years. However fibromyalgia affects you, it is very important to remember that it is a chronic condition and until research has pinned down the cause or causes and developed a cure or at least an effective treatment, you must accept the diagnosis and then move forward with the rest of your life.

What Can I Expect From Treatment?

Studies into treatment have brought some of the most confusing findings. Until doctors know what causes FM, treatment must address the symptoms. Some people respond very well to the standard treatment of low dosage tricyclic antidepressants, aerobic exercise and behavioral modification. Tricyclic antidepressants are given in dosages much lower than that used to treat depression because this family of antidepressant medicines also helps to control pain and improve sleep. Yet many people cannot tolerate even the lowest of dosages of the medicine, and even mild exercise may set off a severe flare-up of symptoms.

> *I have come to view FM as a blessing. It has forced me to slow down and enjoy what is important and let the rest go. No guilt allowed!! I'm a single mom of a 3-year-old and we share life together at an FM pace. —Cindy*

Conventional Western medicine has not been able to bring significant relief for the majority of those with FM and many individuals have turned to alternative medicines. The effectiveness of these treatments is difficult to determine with any real accuracy because controlled studies have either not been done or, in the few cases of studies completed, have not been duplicated. To determine the effectiveness of any treatment, traditional medicine requires that strict control of all factors be maintained. Even one variable can influence the outcome. One of the most common and most difficult factors to control is that of the *placebo effect*. Positive results may be seen in patients who have received a placebo, a pill or regime with no known medicinal effect. The placebo effect occurs in a small percentage of those in a study without any clear reason why, although one thought is that the individual's desire to get better is involved. Alternative medicine looks at the spiritual and mental aspects of health and healthcare, and the placebo effect is one factor being examined.

TWO

The Latest Research

Research on a particular health problem can take a variety of directions. When the health condition is one like FM, researchers can choose to study who gets the condition, what their symptoms are, what kind of changes are found within the individuals' bodies, or how various drugs affect the symptoms and/or those changes within the body. With only 20 years of serious study into the puzzle of fibromyalgia, researchers have made significant strides; however, there is still a great deal to learn and that takes more money and more researchers. In the appendix, we address how you can help researchers but for now, just realize that we still have a long way to go before a cure for FM can be found.

In the meantime, research into the causes of FM and the physical changes it produces, especially within the central nervous system, has produced interesting findings. Lab tests reveal distinct changes in the central nervous system with focus on the neurohormone, endocrine, and immune systems. One question often asked is whether the cause for the pain of FM is in the muscles themselves or within the central nervous system. Significant changes have been found in both.

Research conducted around the world must be compared and confirmed in other studies before we will have a completely accurate picture of the cause of FM. In the meantime, we must try to work within that framework.

We must begin by understanding the biochemical structure and functioning of the human body. Those of us who do not deal with it every day tend to forget that the human body operates on a chemical and electrical

> I research and educate myself, so that I know as much, usually more, than most doctors. I also have a wonderful pillow for my neck pain. It's called Temperpedic (made out of Swedish foam that responds to heat and pressure). I don't wake up with a stiff and achy neck any more!—Tricia

basis. And that all-too-brief look in a high school or college biology class has been buried under more important or at least, obvious, matters.

The very basis of life is dependent upon physical and chemical changes within the body to maintain a stable internal environment. The nervous and endocrine systems work together to maintain this stable environment. The *nervous system* consists of the brain, the spinal cord, nerves throughout the body, and the sense organs. The *endocrine system* is comprised of all the glands that secrete hormones, which are chemical messengers. The endocrine system includes the pituitary, thyroid, parathyroid, and adrenal glands along with the pancreas, ovaries, testes, pineal, and thymus glands. The actions of these glands impact the various systems of the body: the central nervous system, digestive, respiratory, circulatory, lymphatic, urinary and reproductive systems.

A Look at the Muscles in FM

To determine just what happens in the body of someone who has FM, researchers have looked to two separate areas for the source of the pain, the central nervous system and the peripheral areas or muscles themselves. In response to early arguments that fibromyalgia was strictly a psychological problem, scientists looked at the fibers of the muscles. Although there were some minor changes, evidence of microtrauma, there didn't seem to be any physiological changes to explain the pain. Recent studies have become much more complex. While there is little doubt in the minds of leading researchers of FM that the central nervous system plays a very important role in the widespread pain of FM, there is still evidence of changes within the muscles. Researchers have yet to find the cause of these changes, how they come about, and how they can be corrected.

Researchers believe that individuals with FM can be divided into subgroups. What is found, either in cause or influence, for one subgroup may not apply with another. Just as individuals with FM can be broken down into subgroups, so can the evidence of the role muscles play be divided. In early research, despite some appearance of "moth-eaten" muscle fibers, no real abnormality could be found. In more contemporary studies, microscopic abnormalities were found in normal controls as well as in the muscles of FM subjects. It was decided that these changes were normal findings based on the persistent, chronic, biomechanical stresses that act on a muscle.

In 1991, Lindman reported his findings on a series of trapezius muscle biopsies from four groups. The first group was victims in accidental deaths. The second, five healthy females; the third ten women who had work-related trapezius pain; and the last, nine women with FM. Significantly for FM, the capillary endothelial cells were often swollen and deranged in the FM women, while the total capillary area and thickness of the endothelial cells was larger than those in either the normal subjects or the women with trapezius pain. This study demonstrated at the very least that there may be a difference between the muscles of FM and muscles that are either healthy or injured. The clinical importance of this remains to be determined.

Other less invasive studies use nuclear magnetic resonance (NMR) spectroscopy to study the biochemistry of muscles and look at global defects in muscle metabolism. While no evidence of such defects has been found in FM muscles, there are some differences that could occur in calcium-activated muscle damage. Other studies have looked at decreased levels of collagen cross-linking in FM cells that may contribute to changes in and around the cells, including collagen deposition around the nerve fibers. This may contribute to the lower pain threshold found in FM.

Several studies looked at the oxygen levels in the muscles. An early hypothesis that the pain was due to hypoxia (inadequate amounts of available oxygen in the blood), could not be proved. There are some cases, however, that show that while the resting muscle blood flow is normal, the exercising muscle blood flow is significantly lower than that of controls. Because it is the job of the blood to carry oxygen to the cells in muscle and tissues, if the blood flow is lower, there is less oxygen getting to those cells.

A study of muscle tissue oxygen tension found that it was much higher in those with tense muscles than in the normal subjects. The results of this study by Strobel, et al; indicated that hypoxia is not the result of increased muscle tension but from an oversupply of oxygen demanded by the muscle which led to increased capillary flow and rising oxygen tension. The magnetic resonance spectra for inorganic phosphate were higher in patients demonstrating muscle contraction and the intracellular pH was shifted in the alkaline direction with increased muscle tension. If some of these statements seem to contradict each other, it is a common finding in these studies. One study may find that muscle strength is reduced in FM while another says it is normal. One theory is that perhaps there is a growth hormone deficiency which would somehow result in low levels of pro-collagen type III amino terminal peptide, thus reducing muscle strength. In 1997, Norregaard evaluated the physical capacity and effort in patients with fibromyalgia and reported they exhibited significant reduction (about 20 to 30 percent below normal) in voluntary muscle strength of the knee and elbow, flexors and extensors. However, he determined that there was lower effort because of the coefficient of variation being higher. He concluded that there was a low degree of effort yet nearly normal physical capacity.

When my neck and shoulders are hurting but I want to read in bed, I put a pillow on my abdomen and prop my arms on it to more comfortably hold the book.—Barbara K.

When transcutaneous electrical nerve stimulation (TENS) delivered twitches to the muscles during maximal voluntary contraction, 65 percent of the FM patients increased their strength.

Dr. Bennett in his "The Contribution of Muscle to the Generation of Fibromyalgia Symptomatology," which appeared in the *Journal of Musculoskeletal Pain*, stated that "It (the pain) probably arises from focal muscle lesions or a centrally mediated amplification of normal muscle afferent impulses. It seems most likely that both peripheral and central factors play a role in the development of the total FM syndrome."

A study that looked at pressure pain during and following isometric contraction in patients with FM found that there was a decrease in pain threshold that might be due to the sensitization of mechano-receptors caused by muscle ischemia or changes in pain modulation during the contraction. This follows reports from those with FM who say their pain is worse after exercise.

Another look into the relation between muscle pain, muscle activity and muscle co-ordination during static and dynamic motor function found that the maximal voluntary contraction (MVC) during muscle pain was significantly lower than the control condition. At rest there was no evidence of EMG hyperactivity during muscle pain, yet during a static contraction at 80 percent of the pre-pain MVC, muscle pain caused a significant reduction in endurance time. During dynamic contractions, muscle pain resulted in a significant decrease of the EMG activity in the muscle, agonistic to the painful muscle and a significant increase of the EMG activity of the muscle, antagonistic to the painful muscle. According to Graven-Nielsen, Svensson, and Arendt-Nielsen, the muscle pain seems to cause a general protection of painful muscles during both static and dynamic contractions and the increased EMG activity of the muscle antagonistic to the painful muscle is probably a functional adaptation of muscle co-ordination in order to limit movements. This study supports the pain adaptation model which had been postulated by Lund and others that the abnormal distribution of oxygenation in the tenderpoint area was a result of focal area of ischemia.

Myofascial Pain Syndrome

Myofascial pain syndrome (MPS) involves chronic muscle pain that is localized in one area, such as the neck, upper back, or shoulders. MPS is characterized by *trigger points*. These trigger points cause *referred pain* when pressed— that is pain that is felt at a different location. Trigger points have been confused with tender points. They are, however, two different occurrences. One observation that Dr. Bennett made in his article "The Contribution of Muscle to the Generation of Fibromyalgia Symptomatology" in the *Journal of Musculoskeletal Pain* (Vol. 4, No. 1–2, 1996) was that "the muscle origin of FM pain comes from the sequential evolution of an FM syndrome from the focal myofascial pain syndrome." In fact, he quotes one study by Dr. Bengtsson and others that FM "occurs in approximately 80 percent of the patients who suffer myofascial pain syndrome."

An injury to a particular muscle area, such as whiplash injury to the neck, may cause subtle and increasing damage. Over a period of time, the pain from the initial injury spreads to adjacent areas until there is pain in a number of trigger points and soon a regional pain problem develops. If it continues without effective treatment (and sometimes in spite of treatment) it can increase until the widespread pain of fibromyalgia develops. (Figure 2-1).

The effects of a whiplash type of injury is being studied quite closely because of the link to FM. The problem of musculoskeletal pain emanating from the head and neck was addressed extensively in the *Journal of Musculoskeletal Pain*, (Vol. 4, No. 4, 1996). Not all whiplash injuries become chronic pain conditions generally or fibromyalgia specifically, but there have been instances when this was the type of trauma reported as the precipitating stressor. The upper (cervical) spine area and muscles of the neck and upper shoulders have been one of the most prevalent areas of pain in fibromyalgia. A study set in an Israeli occupational clinic found that 21.6 percent of 102 patients developed fibromyalgia following "nonspecific soft tissue injuries to the neck (90 percent having classic whiplash)." Some of the symptoms seen with this type of injury include dizziness, disequilibrium, and the balance problems found in those with fibromyalgia. One of the questions researchers are asking is "What is the link?"

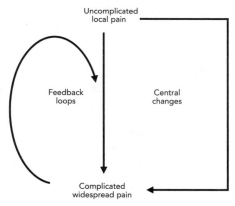

Fig. 2–1. One of the theories on the complexity of FM suggests that numerous feedback loops spread from a local pain problem causing changes in the central nervous system thus developing into the chronic widespread pain. The central pain state or neuroplasticity is the result.

Central Nervous System

Although stress is a word that many individuals with fibromyalgia don't want to hear, there can be no doubt that it does play a role in the syndrome. In the earliest FM research studies, one writer made the statement that "if the individual would eliminate all of the stress in their life, their FM would be cured." Life isn't that simple. Just living involves both good and bad stress; life would be boring without some stress. But the role that stress plays in illness has been studied in increasing detail over the last few years (Figure 2-2).

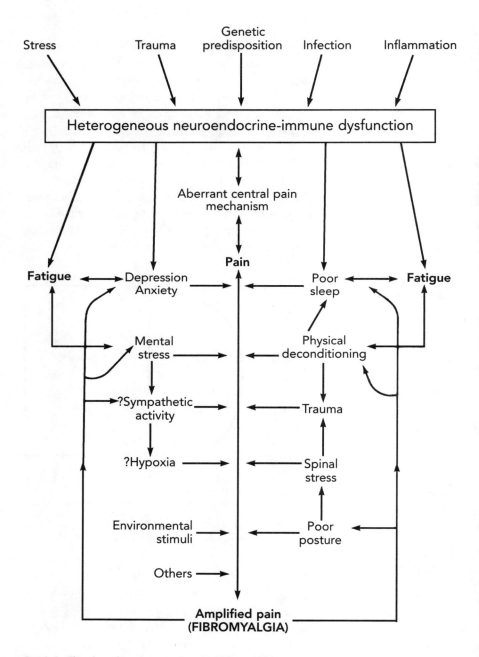

Fig. 2–2. This chart demonstrates the complexity of fibromyalgia and the factors that can impact its development and continuation.

Reprinted by permission, Dr. Muhammad Yunus and *The Journal of Rheumatology* (June 1992, pg. 817).

Most of us are familiar with the "fight or flight response." In primitive times, when danger threatened man, his autonomic nervous system kicked into high gear. The *sympathetic nervous system* prepares the body for action by dilating the pupils, inhibiting salivation, and speeding up the heart rate while shutting down the digestive system. Blood vessels are dilated so increased blood flow and oxygen can be rushed to the muscles. After the danger is past, the *parasympathetic nervous system* slows the body down and resumes the more routine chores of digestion and reproduction.

Researchers looking into the changes that are found in the central nervous system in FM patients are trying to find out whether or not the sympathetic response is turned on but perhaps fails to "turn off," or if the parasympathetic system is impaired when it should be kicking in. Many of the biochemical changes found in study after study include a decrease in *serotonin* levels, one of the chemical neurotransmitters in the brain; increases of up to three times the normal levels of *substance P* in the cerebrospinal fluid; as well as changes in the expected levels of cortisol, epinephrine, norepinephrine, growth hormone, estrogen, progesterone, and thyroid hormones.

All of these substances are involved in one stage or another of the body's stress response, by the *hypothalamus-pituitary-adrenal axis* (HPA). The process is very complex, and some new aspects of it are still being discovered. However, researchers such as Dr. Leslie J. Crofford are trying to determine if an individual's response to stress, particularly chronic stress, is impacted by experiences that occur in the early months of life, genetic factors or something else. Studies in lab rats show that when very young rats are removed from their mothers, their response to normal stressors becomes modified. Could a stressful or abusive childhood lead to FM in adults?

It must be remembered that a stressor is any disturbance in the status quo and can be metabolic, physiologic, traumatic, inflammatory/infectious, or psychoemotional in nature. After stress the sympathetic and parasympathetic nervous systems and the HPA strive to return everything to a steady state. According to Dr. Crofford, these systems must work in unison to regulate themselves. All of these changes and adaptations can occur over a lifetime and be affected by the individual's life experiences. This is another good example of the interrelation of the emotional, mental and physical. Humans do not experience life only on a physical level. Everything that happens to us is processed through our mental and emotional states as well.

(For those who wish to delve more deeply into the subject of stress and FM, Dr. Jay A. Goldstein has written a book, *Betrayal by the Brain,* published by Haworth Press. If you find that book too technical, Katie Courmel has produced *A Companion Volume to Dr. Jay A. Goldstein's Betrayal by the Brain,* which is written in layman's language.)

The HPA axis, and the hormones produced and regulated by it, are also impacted by the natural rhythms of the body. Many FM individuals are known to have disrupted stage four sleep. Many activities of the parasympathetic nervous system are carried on when the body is in this restorative sleep. Are the hormonal and chemical changes found in FM subjects because of the disrupted sleep, or has the parasympathetic nervous system failed to perform as it should and therefore prevented this sleep and the normal biological functions which should be occurring at this time? In looking at the circadian rhythm, many people with FM are out of sync with what is considered normal. For example, it is usual for individuals to experience peaks of energy in the early morning and evening hours with a dip in the mid-afternoon. Yet most people with FM have low levels of energy in the morning and evening, and their time of most energy is late morning to mid-afternoon. Studies into the body's production of such hormones as growth hormone (GH) and cortisol, which has a major role in the HPA axis function, show irregularities in their production and a definite link to the FM patient's unusual internal rhythm.

Not only do individuals with FM report that a stress of some type marked the beginning of their symptoms but studies have shown that, in comparison to other patients with chronic health conditions, those with FM experience a higher level of daily stress or hassles. In keeping with Dr. Muhammed Yunus' *dysregulation spectrum syndrome* theory (see Chapter Three) that many of the associated conditions found in FM sufferers are caused by the same factors, many of these conditions also show abnormalities in the HPA axis function. Abnormal cortisol measurements, in particular, have been most evident in individuals who have had FM symptoms for a long period.

Dr. Crofford and others have raised questions regarding the role gonadal steroids such as estrogen and progesterone may have in fibromyalgia. With 85 to 90 percent of all FM patients female, it would be surprising if there was no connection, but what that connection is remains unclear at this time. Many women report worsening of their symptoms at the time of their menstrual period, and many cases of fibromyalgia flare during the menopausal years.

There are several possible hypotheses regarding HPA stress axis dysfunction and the causes of fibromyalgia but the question remains—do the problems with stress response cause fibromyalgia or are the chemical imbalances found in FM patients a result of the fibromyalgia? One fact has been demonstrated. Of the various treatment strategies tried for fibromyalgia, the most effective work on either recognizing and coping with stress (cognitive behavioral therapy) or on restoring the chemical balance of the body through aerobic exercise or tricyclic antidepressants.

Psychological Studies

People with FM are not simply depressed. Hudson, et al., noted that it would be a disservice to those with FM to label them as having an affective disorder, a psychological problem, or both because this would delay the appropriate medical therapy. Although the medications used for the treatment of FM are often the same as those used for the treatment of depression, the dosage is different. The questionnaires used to assess psychological health often diagnose a person with chronic pain as being depressed, even if that is not their true underlying condition. New research has produced questionnaires specifically geared toward diagnosing FM and these are much more useful in current studies.

There has been speculation that childhood abuse, whether sexual, emotional or physical, is a contributing factor to the eventual development of FM. Two studies, one by Taylor, Trotter, and Csuka in the United States, and another by Boisset-Pioro, Esdaile, and Fitzcharles in Canada, found increased incidences of abuse in the past history of FM patients. The first found 65 percent of the women with FM had been abused sexually, while the figure was 52 percent for the women without FM. They concluded that there is a possibility of a link between the abuse and FM.

The Canadian study reported a high level of sexual abuse in both those with and without FM but they concluded that there is a possibility that abuse might have an effect upon the expression and perpetuation of fibromyalgia in adults.

Carol S. Burckhardt, RN and PhD, is a Professor at the School of Nursing and Assistant Professor of Medicine at the Oregon Health Sciences University and is a member of the multi-disciplinary treatment program there. She is considered one of the leaders in the field of fibromyalgia. In addressing the issue of Abuse and Fibromyalgia in the Fall, 1996 issue of *The Fibromyalgia Times* newsletter for the Fibromyalgia Alliance of America she reviewed the above studies as well as others where researchers found that gastrointestinal patients who had been abused as children were more likely to have abnormal pain perception and be more sensitive to environmental stressors. Burckhardt compared the findings of these studies with the effects stressors have on the HPA axis function and pain perception and indicated that this link may be the real clue to how childhood abuse and certain conditions such as FM and irritable bowel syndrome may be related.

Burckhardt, who writes a column for *The Fibromyalgia Times* called "Life Goes On," urged her readers and those with fibromyalgia to "begin to take back your life, to find the courage to heal, and to know that the effects of any trauma are not always negative." We agree.

However, caution is necessary. When there are two or more conditions which are common in one particular demographic group, care must be taken to prevent faulty assumptions. Just because A and B are both present does not mean that A caused B. Whatever the cause of fibromyalgia eventually turns out to be, the most important thing for someone with FM to do is to move on, striving to be positive in spite of both the cause and the effects of fibromyalgia.

There is one more important fact to note. If depression is present, regardless of its cause, it must be treated along with the fibromyalgia. So too, must any other concomitant health condition that the individual may have as well as FM. This means that one's high blood pressure, diabetes, or heart disease, must be treated in addition to the FM. Just because an individual has been diagnosed with FM doesn't mean that other health problems don't exist. Because so many of the symptoms of FM can mimic other diseases, don't assume everything is caused by FM, don't assume it's not. Use common sense and talk to your doctor about new symptoms.

Pain Perception

Chronic widespread pain and fatigue are the two most obvious symptoms of fibromyalgia. When asked where they hurt, most individuals with FM reply "all over." Early on, it was believed that this was not really the case, that pain was focused on the muscles, tendons, and ligaments, specifically over the tender points, areas where the tendons are attached to the bone. It was the presence of these tender points that helped bring FM out of a psychological catchall into a syndrome with symptoms that were reproducible in patient after patient.

Because the descriptions of pain were so vivid and varied, the severity was often questioned even after tender points were accepted as part of the diagnostic criteria. However, more recent studies have found that when individuals with FM say they hurt all over, they're right. Not only are the tender points painful but so are most of the other areas of the body. Dr. I. Jon Russell, Department of Medicine, University of Texas Health Science Center at San Antonio, discussed the neurochemical causes of FM in the *Journal of Musculoskeletal Pain* (Vol. 4, No. 1-2, 1996). He listed the three clinical findings which have influenced the present views on mechanisms of pain in FM. The three are: the fact that muscles, ligaments, and bursae are all painful and it is unlikely that the same disorder would affect this variety of tissues; the high degree of correlation between the severity of tenderness at the "control points" and the anatomically defined "tender points"; and that dolorimetry (pressure upon the tender points with a small device which delivers a consistent amount of pressure) showed that the individuals had a much lower pain threshold than normal. Many individuals with FM believe that they have a lower pain threshold but also a higher pain tolerance. Dr. Russell states that the higher pain tolerance isn't exactly correct—it's just that FM individuals do endure more pain on a daily basis.

It is important to understand the neurophysiological causes and effects of pain. In order to experience pain, specialized nerve endings called nociceptors are stimulated, transmitting pain messages to the brain for perception. *Allodynia* is the term for a lower than normal pain threshold, and fibromyalgia can then be considered chronic widespread allodynia. With allodynia, sensations that would not ordinarily cause pain do, thus the fact that individuals with FM feel pain from even minor stimuli. *Hyperalgesia* is an overreaction to graded stimuli. Those with FM also report tingling, pins-and-needles feelings, and numbness in their hands and feet. These unpleasant sensations are called *dysesthesia*.

Normally a stimulus of some kind, such as an injury, sends a message to the brain and in response neurotransmitters are released down the spinal cord and into the area of the injury. However, in fibromyalgia there is a "spillover" into the surrounding tissues, increasing the area of pain and making these areas more sensitive to a less intense stimuli (Figure 2-3).

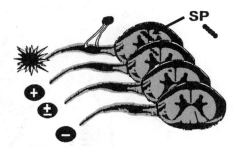

Figure 2–3. Substance P is one of the major neurotransmitters released in the dorsal horm of the spinal cord on activation of muscle nociceptors. Substance P can diffuse over long distances without losing activity and it may be that it does spread into synaptic connections in nearby myotomes so that pain is then felt over a wider area than the initial pain site.

One neurotransmitter, *serotonin*, plays an important role in fibromyalgia and specifically in the pain perception process. In early studies blood levels of serotonin were found to be normal. Analysis of cerebrospinal fluid, however, found a significant decrease in the level of serotonin and a very important increase in the level of substance P. Serotonin works to regulate the amount of substance P, one of several neuropeptides which are involved in the process of pain perception. Vaeroy and others provided an important clue when they showed that substance P in FM patients is elevated at nearly three times the amount found in normal control subjects, and that those amounts are fairly consistent over a long period of time, regardless of the manipulation of the tender points.

Dr. Russell reported that the increased levels of substance P in the cerebrospinal fluid was much higher in FM subjects than amounts taken from individuals who suffered various other painful conditions. The FM subjects in Dr. Russell's study showed that 87 percent of them had levels higher than the highest normal control value. To quote Dr. Russell "The elevated substance P in fibromyalgia is the most dramatically abnormal laboratory measure yet documented in these patients."

Finding these abnormalities may confirm a diagnosis of FM, but does not necessarily mean that the cure is imminent. Dr. Russell went on to state: "Our

1. *Give some time for others. They are so appreciative of your efforts. You can't help but feel better yourself.*
2. *When pillows become flat, put two flat pillows inside one pillowcase to give you the support you need.*
3. *Be your own best friend. Be as kind to yourself as you would for your best friend.*

—Jane

Laboratory Abnormalities in Fibromyalgia

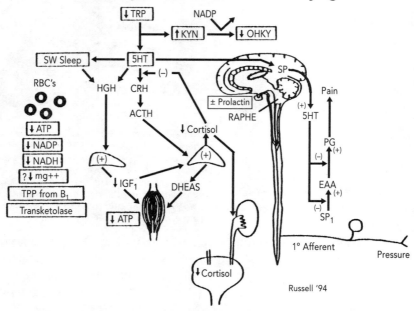

Figure 2–4. This figure shows a summary of abnormalities found in patients with FM. Briefly, they include lower-than normal serum and cerebrospinal fluid (CSF) tryptophan (TRP); increased CSF kynurenine (KYN), decreased CSF 3-hyroxykinurenine (OHKY), lower than normal red blood cell (RBC) adenine necleotides such as adenosine triphosphate (ATP), and a functionally abnormal RBC transketolase enzyme which can be partially corrected by an increased concentration of thiamine pyrophosphate (TPP); decreased productions of hypothalamic/pituitary hormones such as human growth hormone (HGH); decreased production of liver insulin-like growth factor I(IGf1); decreased production of dehydroepiandrosterone sulfate (DHEAS) and cortisol from the adrenal gland; decreased ATP in certain areas of skeletal muscle; and finally, increased substance P (SP) in the spinal fluid. These are the areas of research today.

simplest model to explain these findings would focus on the serotonin deficiency hypothesis... [but] is generally much easier to document an abnormal concentration of a given metabolite in a biological fluid sample than it is to identify the mechanism responsible for that abnormality." See Figure 2-4.

Some researchers feel there may be a link between this increased pain perception and the overresponse to any stimulus that many individuals with FM experience. Some have used the word "hypervigilance" to describe the FM subject's reaction to loud noises, bright lights, strong smells, and even hypersensitivities to drugs, foods, and chemicals. This hypervigilance will be discussed in more depth later.

Research is a continual process. Sometimes by looking for a specific answer to a specific question, the only result is another question. Although the slow research process can be frustrating, nothing can replace a carefully executed, well-thought-out research project. Hopefully, as more physicians become interested in fibromyalgia and more federal and local funding is added to the funds now available, the answers will be on their way.

THREE

Symptoms

Dr. Muhammad Yunus, professor of medicine at the University of Illinois College of Medicine at Peoria has developed a theory he calls Dysregulation Spectrum Syndrome (DSS). DSS presents an umbrella under which fibromyalgia, chronic fatigue syndrome, Gulf War syndrome, irritable bowel syndrome, tension/migraine headaches, primary dysmenorrhea, periodic limb movement, restless legs, temporomandibular pain syndrome, and regional fibromyalgia/myofascial pain syndrome may be grouped (Figure 3-1). There may be more conditions added after more research. His theory states that these have the same cause or causes.

These conditions all share common clinical characteristics. Dr. Yunus' work on the DSS concept dates back to 1981 and his studies on irritable bowel syndrome, tension-type headaches, and migraines. He found these conditions were more common in FM sufferers than in normal controls. Through the years since that early study, Dr. Yunus' work has continued to show links between these conditions. He established the following characteristics to demonstrate the links.

- Clustering of DSS in the same patient groups plus they are more common in any one condition compared to normal controls
- Common symptoms and characteristics: female sex, fatigue, pain, poor sleep, plus a subgroup that shares certain psychological problems like depression, anxiety, and stress
- A generalized hyperalgesic state or a low pain threshold in muscles, tendons, and skin, and also in the viscera
- An absence of "classic" disease: a neurobiological dysfunction rather than a physically abnormal pathology
- Absence of typical psychological disease, yet depression, anxiety, and mental stress occur in a percentage of the individuals, similar to the psychological distress found in other chronic conditions; these do not cause DSS but have an impact on symptoms

Fig. 3–1. Schematic Representation of the Proposed Members of the Dysfunctional Spectrum Syndrome Family

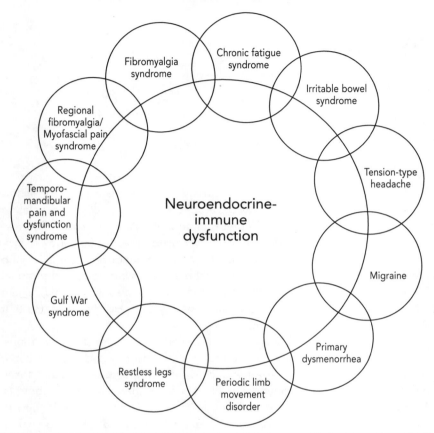

Courtesy Dr. Muhammad B. Yunus and Journal of Musculoskeletal Pain, Vol. 2, #3, 1994, page 14.

- A common genetic factor: so far a genetic link has not been found in FM, but the various conditions grouped together under DSS do tend to show up in families

- Neuroendrocrine-immune dysfunction, including abnormal levels of various neurotransmitters or neurochemicals (serotonin, substance P, noradrenaline, endorphins, dopamine, histamine, and GABA); the HPA axis (hypothalamus-pituitary-adrenal) is also abnormal. Most of these abnormalities are in the central nervous system, cerebrospinal fluid, and brain. Although there are some peripheral factors, these are likely secondary to the central nervous system

- The majority of the patients in the DSS family respond to the same three treatment modalities: tricyclic antidepressants and other such serotonin affecting medications; aerobic exercise; and cognitive behavioral therapy

The importance of dysregulation spectrum syndrome is evident to any-one who has FM and has had to deal with doctors, disability boards, or lawyers. Whatever the triggering factor is, there is a very definite physiologi-cal abnormality present in these common syndromes. If we can get recogni-tion of these abnormalities and therefore these dysfunctional conditions, there will be less money spent on unnecessary tests and referrals to psychia-trists and doctors can focus on finding relief for the pain, fatigue, and other symptoms that affect the individual. In *Fibromyalgia Frontiers* (Fall 1996), the newsletter for the Fibromyalgia Association of Greater Washington, Inc. (FMAGW), Dr. Yunus stated that the cost to this country alone in disability, morbidity and health care for DSS is in the billions of dollars. A recent study by Dr. Frederick Wolfe indicated that the yearly expense of health care for individuals with FM was more than $2,274 a year. Multiplied by an estimated 6 to 10 million Americans, the cost could easily reach the billions, especially when you add associated DSS conditions. A recent Canadian study estimat-ed there were approximately 800,000 Canadians with FM and a separate study estimated the health care for those 800,000 was $400 million (Canadian dollars). Unfortunately, these conditions are still very low on the totem pole of research dollars. Perhaps when a major celebrity or high-profile individual develops FM and decides to actively crusade for research dollars there will be the increase in funds necessary to pin down the cause or causes of FM and DSS and find an effective treatment.

There are many symptoms found in association with fibromyalgia but not everyone with FM has every symptom. Many newly diagnosed individuals look at the list of symptoms and panic. Please don't do that. Although there are some exceptions, generally over the course of time most symptoms are fairly consistent. You may have a few additional symptoms show up after your initial onset of FM, but there is no indication that someone with FM will go on to have all of the symptoms or all of the associated conditions.

Various Symptoms that May Occur with Fibromyalgia

Widespread pain

Fatigue

Sleep disorders

Stiffness

Subjective swelling in fingers

Numbness or tingling in hands/feet

Disequilibrium

Hypermobility

Chest pain

Cognitive Dysfunction – short term memory loss; can't find particu-lar word; poor concentration

Dry, itchy, or blotchy skin

Hand and foot pain (often from repetitive actions)

Hypersensitivities to noises, light, odors

Ocular complaints

When studies show a percentage of FM individuals who may also have particular symptoms or conditions, we will include that information.

We also urge you to take the first step in dealing with fibromyalgia by educating yourself without becoming overwhelmed. Dr. Yunus and others have long advocated that the first step in treating FM is patient education and reading all you can find about FM is part of that first step. I urge you to read carefully and never assume anything. So be safe, be educated and don't assume you are going to experience every symptom or associated condition.

With that stated, let's look at the list of symptoms that have been found in FM.

Widespread Pain

Perhaps the strongest symptom of fibromyalgia is the presence of pain that can be found all over the body and within the body. When the Criteria Committee for the American College of Rheumatology (ACR) examined the records from 16 medical centers across the United States and Canada, they found that widespread pain was present in 97 percent of the cases diagnosed as FM.

Widespread pain tends to occur along the skeleton, in the cervical (neck) area, chest, upper and lower back, shoulders, and hips. It has been described as aching, burning, radiating, spreading, gnawing, and shooting and is sometimes caused or aggravated by muscle spasms. Often it is hard to pinpoint the exact location of the pain, because it can be scattered or diffused. Recent studies have found that the pain can truly be "all over" because of the spillover of the pain messengers into adjacent areas of the skin, tissues, and muscles. Some of the most common sites of pain are the lower back, hips, thighs, abdomen, and shins but many individuals also report pain in the face, chest, and genitalia.

The criteria set forth by the ACR specifies that the pain must be present in all four quadrants (left and right sides, above and below the waist). There is almost always lower back involvement and there is a high percentage of cervical neck pain. It is also necessary that pain be present for more than three months. An acute injury or ailment could be the cause of such pain; but only after it has lasted longer than three months, is it classified as chronic.

For many individuals, the pain begins in one spot and then moves to an overall pattern. This follows the concept that regional myofascial pain that goes untreated spreads out into the nearby tissues and develops into a regional and then widespread pain condition. Not everyone with a myofascial pain condition goes on to develop FM, but there are enough cases of such development that the occurrence is being studied.

When I was serving in the U.S. Navy at Pensacola, Florida, I was required to participate in regular physical fitness tests. The Navy WAVE officers wanted to make the exercises more interesting, so they brought in some of the Navy pilots' physical education instructors, who assigned exercises that placed major stress in my lower back. Everyone experienced sore muscles on the days following the exercises. My situation was more serious because my soreness didn't go away; instead, it got much worse. I had severe pain in my hips and lower back that kept me from taking anything but short, painful steps. It took four months in the hospital with physical therapy to relieve the pain. Through the next few years the primary location of my pain was in my lower back and hips. Only later did it manifest itself in my left chest and shoulders. It is interesting to note that as a child I had twisted and fallen, injuring my lower back. Later, as an adult, I tore some ligaments in my chest when stopping my car suddenly to avoid an accident. Both areas are now sites of most of my pain.

With FM, people experience a general overall aching similar to the aches that accompany the flu. This aching is often present even when the sharper pain is absent. When asked where they hurt, many of those with FM reply, "all over."

Several studies have been conducted to determine the severity of fibromyalgia pain. As in other areas of research, these studies often contradict each other. The pain of fibromyalgia has been compared to rheumatoid arthritis and osteoarthritis, rating much higher than either on visual analog scales (VAS). In one study comparing the pain perception of those with FM to other pain conditions such as that found in diabetic neuropathy or in cases of amputation and "phantom limb" pain, there were higher levels of substance P found in the cerebrospinal fluid of those with FM than in other subjects. Yet, the pain in these other conditions is considered very severe.

There are several ways to measure pain. Several methods were developed in a study conducted by Dr. Sigrid Wigers of Norway and reported in the *Journal of Musculoskeletal Pain* (Vol. 5, No. 2, 1997). The first pain-measurement method had an individual shade all areas, on an illustration of the body, that were painful during a set period of time, such as the last three days. The figures showed both front and back views of the body to give a complete picture. If there was no pain, there was no shading. The total amount of shaded area was converted into a percentage similar to that used for burn victims, i.e., head 9 percent, one arm 9 percent, etc.

The second pain-measurement method was based on pain intensity on the visual analog scale (VAS), with ratings or descriptors ranging from "no pain" to "the worst pain you ever experienced." The researchers also added descriptors such as "some," "medium," and "severe."

The third method of pain measurement checked pressure tenderness, conducted with a hand-held, spring-loaded pressure dolorimeter over the anatomically defined tender spots. A comparison of these three methods showed that the dolorimeter had considerable measurement error while the

VAS was found to be a valid, reliable, and sensitive instrument. However, the use of the patient-made drawings was the best means of determining the patients' own experience of pain modulation.

Because pain is so heavily based on subjective factors, there has been insufficient information in the past to weigh its impact on the individual. However, with the recognition of fibromyalgia and the present acceptance of it as the most prevalent chronic pain condition, there have been improvements in ways of measuring pain impact. Besides an increased level of substance P in cerebrospinal fluid, a report reviewed by Dr. Robert Bennett in the *Journal of Musculoskeletal Pain* (Vol. 5, No. 3, 1997, pg 76-77) showed there were proven changes in the brain after painful stimulation. The readings show that both the medium and long somatosensory action potential "exhibited higher amplitudes and lower thresholds in FMS patients compared to controls." The researchers, J. Lorenz and associates, concluded that "the changes observed in the FMS patients were due to spinal enhancement of impulses or reduced subcortical inhibition of impulses."

According to Bennett "somatosensory action potentials represent electrical recordings in the brain in response to various sensory stimuli." One of the theories about FM has been that there is a state of "hypervigilance," or low tolerance, and threshold to stimuli present. Because this study shows objective evidence of changes which can be explained by an increase in the electrical impulses sent through the spinal cord or by a reduction in the inhibition of the impulses in the brain, it indicates that the problem is an amplification or intensification of the pain signals.

Tender Points

The second strongest symptom of fibromyalgia is the presence of tender points. Dr. Hugh Smythe first suggested that tender points be used as part of the diagnosis, suggesting that 12 of 14 such points be a positive sign. During the late 1970s and 1980s researchers and doctors increased the number of tender point locations to 75, using various percentages to indicate a positive diagnosis. There was no clear standard established until the Criteria Committee met in 1989.

The Criteria Committee decided on 18 sites, located on both sides of the body and in areas above and below the waist. In a study conducted by the committee, the presence of the tender points was the most powerful indicator separating those with fibromyalgia from those without. In the control, or normal, group, subjects felt a slight tenderness but no actual pain when pressure was applied to a tender point.

But when that same amount of pressure was applied to a tender point on a person with fibromyalgia, the examiner got an intense pain response, with

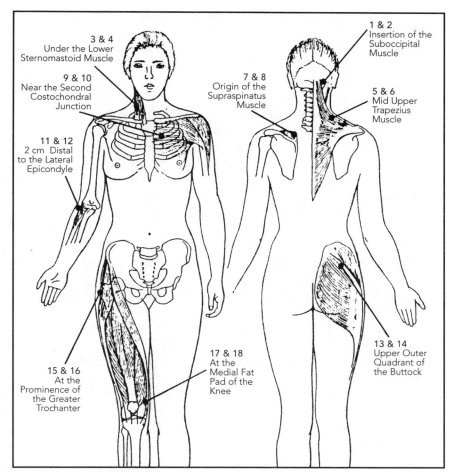

Fig. 3–2. Tender Point Sites

the person often pulling away. In early studies, doctors looked for this so-called jump reflex as a pointer to FM.

There are several reasons that these tender points are important, particularly to a person with FM. Because of the early tendency to attribute the pain and fatigue to psychological problems, the discovery of these sites of pain was a great relief to both patient and doctor. (The medical community endorses the tender-point theory because a person being tested has no knowledge of such points and therefore cannot fake pain at that point.)

Tender points are in specific locations. They are situated at the bony insertions of tendons or at a muscle-bone interface. If the examiner is off even centimeters from the indicated spot, the response is not the same. Recent studies suggest, however, that even nearby areas can be painful.

In addition to relying on the presence of tender points to disprove a mental illness diagnosis, we also know that psychogenic rheumatism, which is a manifestation of a psychological problem, produces erratic and bizarre symptoms, none of which can be reproduced by an examiner.

Because it is often difficult to apply a consistent amount of pressure, many doctors use a simple spring-loaded gauge, called a dolorimeter, which enables them to be more precise in measuring the amount of pressure they applied. The committee determined that pressure must be applied with an approximate force of 4 kg. They also determined that at least 11 of the 18 points should register positive for the diagnosis of FM (Figure 3–2). The 18 sites are the following:

- Sites 1 and 2, occiput: bilateral, at the suboccipital muscle insertions
- Sites 3 and 4, low cervical: bilateral, at the anterior aspects of the inter-transverse spaces at C5-C7
- Sites 5 and 6, trapezius: bilateral, at the midpoint of the upper border
- Sites 7 and 8, supraspinatus: bilateral, at origins, above the spine of the scapula near the medial border
- Sites 9 and 10, second rib: bilateral, at the second costochondral junctions, just lateral to the junctions on upper surfaces
- Sites 11 and 12, lateral epicondyle: bilateral, 2 cm distal to the epicondyles
- Sites 13 and 14, gluteal: bilateral, in upper quandrants of buttocks in anterior fold of muscle
- Sites 15 and 16, greater trochanter: bilateral, posterior to the trochanteric prominence
- Sites 17 and 18, knee: bilateral, at the medial fat pad proximal to the joint line

Fatigue

"I've had a total reversal of lifestyle. I was working full-time at a challenging, good-paying job. Recreational activities included hiking in the mountains, swimming 120 laps a day in a 25-meter pool, aerobic exercise and dancercize."

Fatigue presents almost as much of a problem for the person with FM as pain does. Quite often it is possible to concentrate on something else, to take your mind off the pain much of the time, but fatigue can leave you without the ability to even concentrate. Fatigue is present in nearly 90 percent of those seeking medical care and may be one of the primary reasons for seeing a doctor. Such overwhelming fatigue is likely to be caused by a central neurohormonal mechanism, a fact which supports a more complex cause than just "deconditioning." Dr. Leslie Crofford, assistant professor of Internal Medicine, Division of Rheumatology, University of Michigan, probably the leading researcher of the hypothalamic-pituitary-adrenal (HPA) stress axis in FM, indicated that the fatigue of FM may be related to abnormalities of the HPA axis. She stated "Subtle adrenal insufficiency states have been proposed to be a common denominator in a number of chronic fatigue states, with central corticotrophin-releasing hormone (CRH) deficiency postulated to be one mechanism underlying this type of insufficiency." CRH is the dominant hypothalamic mediator or stimulus of HPA axis activation.

The effects of fibromyalgia fatigue vary from person to person; some people are only aware of a tired feeling after exertion. These people may need no more than a short nap in the afternoon to continue their normal activities. For others, however, even minor exertion produces extreme fatigue, requiring them to rest often.

People with FM often find that short periods of exertion require much longer periods of rest to recuperate. One person says that for a half hour of exertion, she needs at least a full hour to recover. As a result, it takes much longer to complete tasks because frequent rest stops are necessary. One of the most common complaints from people with FM is that they are no longer able to do even normal daily tasks. They worry about returning to full- or even part-time work when they need help with the basic activities necessary to maintain day-to-day life.

Sometimes shortness of breath occurs and some physicians have speculated that the fatigue, and hence, the shortness of breath, is related to being out of shape or "deconditioned." Deconditioning occurs when an individual does not exercise and therefore use muscles regularly, leaving the muscles open to cramping and pain when they are used. The link between conditioning and restorative sleep was found when Moldolfsky studied the effects of waking subjects during the Stage Four or Non-REM (Rapid Eye Movement) Sleep. It is at this level of sleep that the body is refreshed. Those subjects who regularly participated in aerobic exercise, such as cross-country running, did not develop the muscle pains and other symptoms that subjects who did not exercise regularly did. Often, individuals with FM will stop exercising or being as active because of muscle pain, which increases with activity.

More recent studies have shown that people with FM can exercise to exhaustion under controlled conditions. They have confirmed what many individuals have stated; that is, that they hurt more after exercise. The studies indicate the likely problem is one of the central nervous system rather than one involving local muscles. So it is more than a matter of just being deconditioned.

A number of recent studies have looked at the role muscles and aerobic fitness play in FM. Some research indicates that if a person with FM works to become aerobically fit, the FM symptoms will ease. Aerobic exercise is one of the cornerstones of treatment for FM but it must be begun slowly and increased slowly or pain and fatigue can increase significantly. Anyone beginning an aerobic exercise program should discuss it with their doctor or physical therapist and start out with stretching exercises to warm up.

Sleep Disorders

If the focus of the question is on the ability to fall asleep and stay asleep, many people with FM say they don't have a problem when questioned about their sleep habits. However, if asked how they feel when they get up in the morning, most will say they feel terrible. Others describe themselves as light sleepers who feel terrible in the morning. A now-cliched saying that is still nonetheless true, is "I wake up feeling like I've been run over by a Mack truck."

No matter how many hours of sleep they get, almost all people with FM awake unrested. Dr. Moldofsky has been studying the sleep problems associated with FM since the 1970s. His research has demonstrated an intrusion of *alpha*, or waking brain waves, into the *delta* waves of stage four or non-rapid-eye-movement (non-REM) sleep, the stage of sleep that restores the body's energy. This is called *alpha-delta sleep*. When Dr. Moldofsky continually interrupted the stage four sleep of healthy subjects, they developed the fatigue, musculoskeletal pain, and tender points typical of fibromyalgia.

Fig. 3–3
Factors Important in Suspecting Sleep Disturbance

- Difficulty in initiating sleep
- Difficulty maintaining sleep
- Fatigue on awakening
- Pain and stiffness on awakening
- Excessive sleepiness or napping during the day or evening
- Vigorous exercise in the evening
- Excessive fluids after dinner
- Excessive caffeine
- Excessive alcohol
- Excessive heavy or spicy foods in the evening
- Watching television before bedtime

- Eating before bedtime
- Nocturia (the need to urinate at night)
- Excessive noise
- Uncomfortable mattress
- Poorly controlled temperature
- Sleep medication
- Periodic leg movements
- Restlessness
- Snoring
- Pets in the bedroom or on the bed
- Excessive stress

Courtesy of Dr. P. Kahler Hench. Reprinted with permission from *Rheumatic Disease Clinics of North America*.

New studies have found that the intrusion of alpha waves not only is present in the delta, or deep, sleep, but also can occur in stages one, two and rarely in rapid-eye-movement (REM) sleep. The new term for this, according to Dr. Moldofsky, *is alpha EEG (electroencephalographic) NREM sleep anomaly*. Other studies have shown that while normal controls of those with chronic insomnia (dysthymia) may have approximately 25 percent of their NREM sleep interrupted with alpha wave intrusions, those with fibromyalgia have about 60 percent of their NREM sleep interrupted. Interestingly, there are individuals who have the alpha-delta intrusion but do not have the symptoms of fibromyalgia. This has caused doctors and researchers to wonder if there are certain individuals who may be predisposed to the alpha-delta condition and develop FM only after it is triggered by an accident, illness, or stress-related state of crisis.

These intrusions of alpha waves into the delta, or deep, sleep waves have been compared to an internal arousal mechanism that kicks in on a regular basis. It may be this mechanism that pulls sleepers out of their restorative sleep, leaving them tired and in pain. The level of pain seems to be directly correlated to the length of time that delta-wave sleep was disrupted. According to a report published in *Pain*, (Dec. 1996), there appeared to be a "significant bidirectional within-person association between pain attention and sleep quality that was not explained by changes in pain intensity." Other studies have found a repetitive cycle of poor sleep, increased pain and fatigue leading to more poor quality sleep.

Another study examined the relationship between alpha sleep and information processing during sleep, perception of sleep, musculoskeletal pain, and arousability in patients with FM. The findings indicated that alpha sleep was, electrophysiologically, a shallow form of sleep and that it was associated with increased arousability—thus proving the earlier thesis that alpha interruptions into deep sleep may cause poor sleep quality. Still another study compared self-reported sleep quality and psychological distress as well as somnographic sleep and physiological stress arousal. The women with FM reported poorer sleep quality and higher levels of psychological distress. Interestingly enough, this study did not show the expected alpha EEG-NREM activity.

Patients in a sleep disorder clinic were tested for fibromyalgia in another study with no positive matches. This is an indication that the sleep disorder did not come first, but was probably a result of another factor in the equation. Another study on people with such sleep disorders as sleep apnea (obstructed and irregular breathing during sleep) and sleep-related periodic limb movement syndrome (PLMS) or restless leg syndrome showed these patients to comprise a higher percentage of men than women. It was suggested in this study that the nonrestorative sleep found in fibromyalgia could be an effect of the FM (or the initial trigger) and not the cause. It also con-

---

Fig. 3–4
Factors Physicians Have Found to Have a Negative Effect on the Sleep of People with Fibromyalgia

• Intrusion of alpha waves into stage-four delta-wave sleep

• Bruxism (grinding teeth during sleep)

• Nocturnal myoclonus (periodic leg movements)

• Sleep apnea (temporary cessation of breathing)

• Narcolepsy (extreme tendency to fall asleep whenever in a quiet environment)

• Nocturnal hypnogenic myoclonus (nighttime hypnotic state with periodic leg movements)

• Arthritic pain (rheumatoid arthritis and osteoarthritis)

• Stress

• Symptoms of concomitant illness

• Nocturia (passage of urine at night)

• Pain

Courtesy of Dr. P. Kahler Hench. Reprinted with permission from *Rheumatic Disease Clinics of North America.*

---

cluded that "PLMS and K/alpha sleep probably don't represent different manifestations of an identical central nervous system process," according to Dr. Robert Bennett who reviewed the literature. K/alpha sleep refers to an anomaly on an electroencephalographic (EEG) or brain wave study, which occurs periodically (every 20-40 seconds) and is then followed immediately by alpha-EEG activity for one-half to five second duration. Other publications state the rate of restless leg syndrome in those with FM to be as high as 50 percent and because the ratio of women to men with FM is so high, there is bound to be a large group of women who have a restless leg or periodic leg movement problem. Again, the question arises, which came first, the symptom or the syndrome? It should be noted that another study on sleep apnea found a significantly higher number of men than women to also have the sleep disturbance. Sleep apnea is a group of sleep disorders in which the sleeping person repeatedly stops breathing long enough to decrease the amount of oxygen in the blood and brain and to increase the amount of carbon dioxide. Because of the possible serious consequences of leaving sleep apnea untreated (high blood pressure and cardiac problems), individuals who do have it should seek treatment (See Figure 3–4).

When trying to determine what causes the alpha-EEG-NREM sleep anomaly, a number of factors have been considered, including psychological stress, environmental events, primary sleep and physiology disorders, and painful joint problems. There have been some interesting findings in these most recent research projects. A paper presented by Drs. Sutton, Moldofsky, and Badley, for the Arthritis Community Research and Evaluation Unit, Wellesley Hospital Research Institute, Toronto, Canada, and delivered at the 1997 Annual Meeting of the American College of Rheumatology dealt with the ties between arthritis, pain, chronic conditions, and stress as being determinants of insomnia and poor sleep quality in the population. They found that "the presence of arthritis is an important independent determinant of sleep difficulty in the population, after adjusting for *pain, long term disability, other chronic conditions, and stress.*" (My emphasis) In other words a chronic pain condition can cause sleep disturbance resulting in nonrestorative sleep.

Ongoing research into the link between sleep disturbance and FM includes comparing people who experience a noninjury automobile or industrial accident and develop sleep disturbances and musculoskeletal pain to those who have no known cause for their fibromyalgia symptoms. Those who had an accident or experienced domestic problems from which they could not extricate themselves showed much stronger sleep anomalies.

Problems with the sleep environment have also been proven to cause disruption. Many people with fibromyalgia have indicated a sensitivity to noise that may explain their sleep problems. Studies conducted near airports have shown that residents report sleep problems, fatigue, rheumatic pain, and emotional distress. Other studies suggest that the wrong bedroom temperature, the presence of young children, having pets in the room or on the bed, and even sharing the bed with someone who snores may disrupt sleep.

In recent studies into neurohormonal disturbances associated with FM, the importance of sleep has been shown in the production of serotonin and substance P, as well as immune and neuroendocrine substances such as interleukin-1, growth hormone, and cortisol.

New research is also looking into diurnal aspects of FM. Fibromyalgia somehow affects the internal body clock. In those without FM, early morning and evening are the most alert times, with a dip in energy levels in the early afternoon. However, this is reversed in those with fibromyalgia. Upon awakening, those with FM are slow to get around and feel badly. They report their best time is early afternoon but, by evening, fatigue and pain have risen and they feel worse. More studies are continuing into the role of the body's natural rhythm, the role the HPA axis function plays in it, and in sleep.

Morning Stiffness

"It takes me nearly an hour to work out my stiffness when I get up in the morning."

Nearly 80 percent of individuals questioned report morning stiffness, and there has been a direct correlation shown between the sites of pain and this stiffness. While more studies are looking at a possible central nervous system cause, some activities may have an impact on stiffness as well as inactivity.

The morning stiffness of fibromyalgia has been compared to that experienced by people with rheumatoid arthritis, although there is not yet a definitive explanation. One theory suggests a physiological problem may be the answer. As Dr. Xavier Caro explained in the *Fibromyalgia Network Newsletter* (April 1990), "It is due to the accumulation of tissue fluids and mediated, at least in part, by a lesion of increased vascular permeability. In other words, proteins and other substances are able to leak through the blood vessel wall barriers (due to these small vascular lesions) and accumulate in the tissue areas where they are not supposed to be."

Some people report a staggering or stumbling walk when they first get out of bed in the morning. This may be caused by the stiffness and eases as the person walks around. During the day, keeping a limb or hand in one position or sitting for a prolonged time may cause the stiffness to return. Stiffness can also occur during weather changes.

Other Symptoms

- Subjective swelling in the hands or feet, evident only to the individual.

- Paresthesia or dysesthesia, or sensations of numbness or pins-and-needles prickling in the hands or feet.

- Disequilibrium or a sensation of dizziness or vertigo. There may be a link to orthostatic sympathetic derangement or neurally mediated hypotension which produces a drop in the blood pressure after a person has been lying down. Unlike orthostatic hypotension which creates the sensation of dizziness or faintness upon arising, neurally mediated hypotension can occur after some time has passed. Tilt table testing has proven that there is a strong link between this significant drop in blood pressure, which indicates an improper response of the sympathetic nervous system, and fibromyalgia and chronic fatigue. One study in Sweden found 74 percent of those with FM exhibited dizziness or other signs of disequilibrium. For most of them, the symptoms were mild, but for nearly a quarter, the symptoms were severe.

- Hypermobility or loose joints, usually exhibited as a turned ankle or knee or even fingers or the wrist "giving way" to weight or a strain. This contributes to stumbling or falling as well as dropping items. One recent study reported no findings of hypermobility in FM patients while others have found numerous examples. This may be another instance of subgroups among those with FM.

- Chest pain, sometimes caused by *costochondralgia* or *costochondritis* pain in the muscle at the spot where the ribs meet the chest bone. Other times it may be a referred pain or a non-cardiac chest pain which may mimic a heart problem.

- Sicca symptoms or dry eyes and dry mouth from reduced production of tears and saliva. This may be made worse by certain medications commonly used to treat FM.

- Skin problems, especially dry, itchy, or blotchy skin.

- Hand and foot pain (may also be carpal tunnel syndrome) often occurs with repetitive motions, such as typing, use of a computer mouse, driving a standard shift car for the first time, or after driving an automatic. Some foot pain may also be due to plantar warts; high- or low-arched feet, which throw the toes out of position; bursitis of the Achilles tendon; or heel spurs.

- Hypersensitivities to noises, light, and odors were originally described as being "irritable to everything" but recent research may point to an abnormal response to stimuli. Hypersensitivity can also be described as multiple chemical sensitivity (MCS), a chronic condition that produces increased response to multiple, different chemicals and other irritants at or below levels which they were previously tolerated. This is not a true allergy although many individuals call it that. Examples of stimuli range from food and drugs to lights, noises, and odors. ("Sick building syndrome" could be included.) Individuals with MCS can have a sudden onset of symptoms or a more gradual one. Symptoms can include many of those found in FM—fatigue, headache, poor sleep, etc.

- Ocular complaints; besides dry eyes, some individuals report problems with focusing as well as difficulty in tracking movement.

- Auditory problems like low frequency, sensorineural hearing loss (decreased hearing due to damage or disease of the auditory nerve); decreased painful sound threshold; occasional noises or ringing in the ear (tinnitus). Studies don't agree on hearing loss but an auditory brainstem response (ABR) test which gives information on the cochlea, auditory nerve, and the brainstem, may show significant changes which indicate brainstem dysfunction in the FM subject.

- Cognitive dysfunctions like short-term memory loss, poor concentration, forgetfulness, and confusion can affect up to two-thirds of those with FM. Some medications make the problems worse as can fatigue, mood disturbances, pain, and sleep disturbances. Some studies have shown a direct correlation between poor sleep and cognitive functioning. Early research said little about this "fibro-fog" as it has been dubbed, but recent studies have demonstrated clear evidence that the problems do exist. Single-photon-emission-tomography (SPECT) scans of the brain have shown decreased blood flow in the caudate nuclei and the thalami in FM patients. These are important areas of involvement for those with FM. The thalami is involved with sensory messages going to and from the highest levels of the brain, and is also responsible for short term memory which is easily disturbed at this level. The caudate nuclei is part of the basal ganglia which is still not completely understood. However, it is known that the neuron cell bodies within the caudate nuclei (and the putamen and globus pallidus, also part of the basal ganglia) serve as relay stations for motor impulses originating in the cerebral cortex and passing on into the brain stem and spinal cord. These impulses also aid in controlling various muscular activities.

As newer and more specialized research studies are conducted and more complex tests are done, more symptoms may appear linked to FM. Remember again, not everyone will have every symptom. It is most likely that new research will help to divide those with FM into subgroups, perhaps based on which symptoms they exhibit, and from there, more effective treatments will be found.

A point to remember is that you can't automatically assume that every symptom or health problem you develop is caused by FM. Talk to your doctor and use your own common sense and awareness of your body to determine if a new problem requires looking into. Don't let FM-like symptoms mask other, more serious health problems.

FOUR

Associated Conditions

Dr. Yunus is one of a growing number of researchers who believe that the linked symptoms of FM and DSS are more than just chance or coincidence. It is possible that when researchers start to find out how to separate those with fibromyalgia into subgroups, the presence of some of these conditions may be a determining factor. It is important to remember that, just as with the symptoms covered in the last chapter, not everyone will have every one of these conditions. We will also include some conditions which show up in those with FM but may have no link with other associated conditions. We will look at some of the symptoms that the various syndromes share as well as some they don't.

Associated Conditions Seen with Fibromyalgia

Chronic fatigue syndrome (CFS)

Irritable bowel syndrome

Irritable bladder/interstitial cystitis

Heartburn/GERD and/or esophageal dysmotility

Gulf War syndrome

Multiple chemical sensitivities (MCS)

Migraine/tension headaches

Dysmenorrhea/vulvar/vaginal pain

Thyroid disease

Temporomandibular Joint Dysfunction (TMD)

Mitral valve prolapse

Depression/anxiety

Increased daily hassles

Regional fibromyalgia/myofascial pain syndrome(MPS)

Craniovertebral stenosis

Illnesses that sometimes precede FM:

Lyme disease

Hepatitis C

Silicone breast implants

HIV

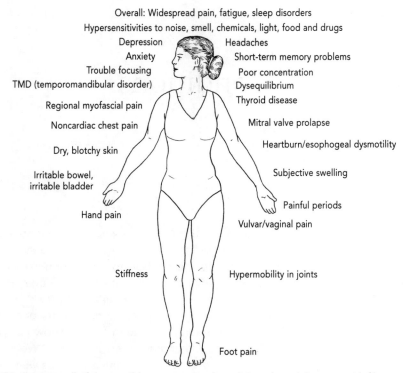

Overall: Widespread pain, fatigue, sleep disorders
Hypersensitivities to noise, smell, chemicals, light, food and drugs
Depression — Headaches
Anxiety — Short-term memory problems
Trouble focusing — Poor concentration
TMD (temporomandibular disorder) — Dysequilibrium
— Thyroid disease
Regional myofascial pain
Noncardiac chest pain — Mitral valve prolapse
— Heartburn/esophogeal dysmotility
Dry, blotchy skin
Irritable bowel, — Subjective swelling
irritable bladder
— Painful periods
Hand pain
Vulvar/vaginal pain
Stiffness — Hypermobility in joints
Foot pain

This illustrates all of the possible symptoms and conditions that might occur with fibromyalgia.

Chronic Fatigue Syndrome (CFS)

Perhaps the most common condition that is found linked to fibromyalgia is *chronic fatigue syndrome* (CFS) or also called *chronic fatigue immune dysfunction syndrome* (CFIDS). It is sometimes thought that fibromyalgia and chronic fatigue syndrome are the same condition, but, though they share some common symptoms, there are some major differences. Generally, the strongest symptom with CFS is chronic fatigue while those whose major complaint is musculoskeletal pain are diagnosed as FM. Both conditions have had to fight the tendency of many to label them psychological; the phrase "It's all in your head," is enough to send individuals with either condition into a fury. Dr. Marsha Wallace, an internist in private practice in Washington, D.C. and an assistant professor of medicine at George Washington University, treats both conditions. She wrote a column comparing them in *Fibromyalgia Frontiers* (Fall 1996), a newsletter for the Fibromyalgia Association of Greater Washington, and stated that "The fact is both fibromyalgia and chronic fatigue syndrome *are* 'all in your head,' but more specifically, they are in your brain, not your imagination. Those doctors who have been telling you that it's in your head did not know how close they were to the truth!" She went on to say "Fibromyalgia and chronic fatigue may both be the result of brain damage, infection, or stress. It is not clear at this point."

Diagnostic criteria for CFS were established in 1988, two years before those for FM. Both will probably have to be fine-tuned before they can truly define these syndromes. (The criteria for CFS were revised in 1994.) Although FM can be diagnosed when another chronic illness such as rheumatoid arthritis or lupus is present, CFS cannot. There cannot be another illness or condition that might be causing the fatigue. (This does not include FM, anxiety disorders, somatoform disorders, nonpsychotic or nonmelancholic depression, neurasthenia or multiple chemical sensitivity disorder.)

Diagnostic criteria for CFS states: "Chronic fatigue syndrome is a clinically evaluated, unexplained, persistent, relapsing chronic fatigue that is of new or definite onset, that has not been lifelong. It is not substantially alleviated by rest and results in substantial reduction of previous levels of activity—occupational, educational, and social." There is no longer a set length of time for the symptoms to have been present.

Individuals must also have four of the following eight symptoms for six months:

- Self-reported impairment in short-term memory or concentration severe enough to cause substantial reduction in previous levels of occupational, educational, social, and personal activities
- Sore throat (without infection)
- Tender (not enlarged) cervical or axillary lymph nodes
- Muscle pain
- Multi-joint pain without joint swelling or redness (therefore, not real arthritis)
- Headaches of a new type, pattern or severity
- Unrefreshing sleep
- Post-exertional malaise lasting more than 24 hours

FM and CFS share several characteristics such as a triggering factor or stressor, whether it was an infection, trauma, surgery, or an emotional stress. Researchers for both syndromes are looking into the role of the central nervous system in three areas: the autonomic nervous system, the hypothalamic-pituitary-axis, and the nociceptive pathways. The problems with these systems may not be the same in each syndrome but there are enough links to tie the conditions together.

There are two to four times more women with CFS than men, and while there is no standard treatment protocol, some trials with tricyclic antidepressants such as are used with FM, some antiviral drugs and immunomodulators (drugs that boost the immune system) have been found to provide some improvement. As in FM, treatment of the symptoms is the primary focus.

Irritable Bowel

Irritable bowel syndrome (IBS) is one of the more common conditions seen with FM. One recent study found it in 42 percent of FM patients but in only 16 percent of controls. In IBS, the individual may have constipation, diarrhea, or an alternating cycle between the two. Usually there is abdominal pain, often relieved by bowel movements; bloating of the abdomen, mucus in the feces; and excessive gas. IBS is sometimes also known as irritable colon syndrome and spastic colon. Like the conditions in this chapter, the cause is not known but again the link to the central nervous system is a major area of research. The symptoms of IBS come and go just as do those of FM but they are definitely exacerbated by stress. A report from the Medical World News, April 1988, found on the Sapient Health Network web site, indicates that two independent studies linked IBS to a "hypersensitivity of a patient's intestinal nervous system." This "hyperresponsiveness can be traced to the enteric nervous system that enervates the intestine." This is another instance of an individual's hypersensitivity in response to a stimulus.

According to the American Medical Association's *Home Medical Encyclopedia*, "the basic abnormality is a disturbance of involuntary muscle movement in the large intestine." It is the most common intestinal disorder, accounting for more than half of all visits to gastroenterologists, and is found more often in women than in men. The diagnosis is made by ruling out any physical causes such as infection, endometriosis, ulcerative colitis, or colon cancer. A thorough medical history and a physical exam, routine blood test, perhaps a lactose intolerance test, and a stool sample are used to complete the diagnosis. If necessary, the doctor may request a sigmoidoscopy, which allows the lower bowel to be viewed directly.

Treatment usually is directed toward the diet, because IBS is also aggravated by eating certain foods. Bulk-forming agents such as bran or methylcellulose may be recommended for constipation while an antispasmodic drug may be prescribed. In the case of diarrhea, both over-the-counter and prescribed antidiarrheal drugs may be used. The patient may be told to try a period of avoiding dairy products in case they aggravate the problem.

Irritable bowel syndrome can have a major impact on an individual's daily activities, particularly if diarrhea is the primary symptom. For others, it may only be a minor inconvenience. Some women report that just as their FM symptoms tend to worsen at the time of their menstrual periods, so do the symptoms of IBS.

Stress is considered a factor in IBS, more easily understood when you realize that bowel upset is a common reaction to stress in normal, healthy individuals. When researchers began to look at the relationship between FM and IBS, some studies showed that irritable bowel symptoms were present before fibromyalgia symptoms appeared.

Irritable Bladder/Interstitial Cystitis

In the late 1980s and early 1990s, some people were calling those individuals with FM "Irritable Everything," referring to IBS plus some of the other hypersensitive responses that include the bladder. Individuals with FM often find themselves making frequent trips to the bathroom, even having to get up at night several times to urinate, yet there is no evidence of a bladder infection. There may be pain with urination or suprapubic pain. Some doctors believe that *irritable bladder* and *interstitial cystitis* are two separate conditions while others use the terms to mean the same condition. Regardless of the term used however, there is increased urinary frequency and urgency, and lab tests fail to detect any bacteria that might be causing the problem. In a recent study done by Dr. Daniel Clauw, there is a clear link between interstitial cystitis and other chronic conditions of the DSS group, showing similar demographics, natural history, aggravating factors, and efficacious therapy plus the presence of irritable bowel, migraine, and increased pain perception.

In another study by Dr. M. Alagiri published in *Urology* (May, 1997), researchers found that those with interstitial cystitis were 100 times more likely to have irritable bowel syndrome than the general population and 30 times more likely to have lupus, a rheumatic illness often found in those who have FM. This study also found that irritable bladder and irritable bowel have several other dysfunctional conditions in common.

Diagnosis of either condition is based upon medical history, urine studies, and a pelvic exam. A cystoscopy, which looks inside the bladder, is also done to rule out other possible causes. Both irritable bladder and interstitial cystitis may benefit from increased intake of water. The individual may also try eliminating certain foods and beverages that aggravate the problem, such as caffeine in coffee or soft drinks, carbonated drinks, and spicy or acidic foods. Tricyclic antidepressants such as amitriptylline are a common prescription as well as phenazopyridine hydrochloride, antihistamines, or calcium channel blockers. Some medications that are placed directly in the bladder may be used such as dimethyl sulfoxide, steroids, or heparin. Some of these must be inserted by catheter under either a regional or general anesthetic. Many women find that as with IBS and FM, their symptoms may be worse around menstruation or menopause and hormone therapy may help. Stress management is also a possibility.

Heartburn/Gastroesophageal Reflux Disease/ Esophageal Dysmotility

Heartburn is one of the symptoms of *gastroesophageal reflux disease* (GERD) as are chest pain and regurgitation of gastric juices and some small amounts of food from the stomach. Although there has not been very much written about the relationship between FM and GERD or heartburn (which can occur without the presence of GERD), Dr. Daniel Clauw listed problems with reflex swallowing abnormalities in smooth muscle functioning and tone in the esophagus, and heartburn in "New Insights Into Fibromyalgia" in *Fibromyalgia Frontiers* (Vol. 2, No. 4, Fall 1994), in a list of symptoms and syndromes possibly linked to fibromyalgia.

Gulf War Syndrome (GWS)

Increasing research indicates that many symptoms that veterans of the Persian Gulf War have exhibited since their return from that war match those shared by fibromyalgia, chronic fatigue and the other conditions Yunus included in his dysregulation spectrum syndrome (DSS). More than 996,000

Listed below are research projects that may benefit those with FM

Brain & Nervous System
- Dysregulation of the stress response in the Persian Gulf syndrome
- Evaluation of cognitive functioning of PGV
- Evaluation of neurological functioning in PGV
- Neuropsychological functioning in Persian Gulf era veterans
- Psychosocial, neuropsychological and neurobehavioral Assessment (Project I)
- Fatigue in Persian Gulf Syndrome—physiologic mechanisms
- Evaluation of muscle function in PGV
- Psychological and neurobiological consequences of the Gulf War experience
- Evaluation of cognitive functioning in PGWV reporting war-related health problems
- Neuropsychological functioning in veterans
- Female gender and other potential predictors of functional health status among PGWV

Immunology
- Clinical and neuroendocrine aspects of fibromyalgia (Project II)
- Immunological evaluation of PGV
- Musculoskeletal symptoms in GWS
- Chronic gastrointestinal illness in PGV
- Diarrhea in PGV: an irritable bowel-like disorder
- Effects of exertion and chemical stress on PGV
- Physiological and psychological assessments of PGV

PGV – Persian Gulf Veteran, GWS – Gulf War syndrome, PGWS – Persian Gulf War veterans

---

servicemen and women served in the Gulf from August 1990 through the end of 1995. While the government initially denied any exposure to potentially harmful biochemical weapons or agents, it has since released information that some of those serving may have been exposed to small amounts of sarin, a nerve agent. As many as 98,900 U.S. troops were exposed when the troops demolished a munitions depot at Khamisiyah in southern Iraq. Other possible causes of the symptoms reported include the burning oil wells of Kuwait; the doses of a nerve agent pretreatment drug, pyridostigmine bromide, given to the troops in case of biochemical warfare; exposure to the rubble and dust from exploded shells made from depleted uranium or from handling those shells; or some as yet unknown Iraqi chemical-biochemical agent. What is known, however, is that those service men and women faced a great deal of stress, not only the usual stress that combat brings, but also from the very uncertainty about the chemical weapons that Iraq had and might use.

Symptoms exhibited by the veterans include fatigue, headache, muscle pain, joint pain, rash, cognitive complaints, sleep disturbance, GI complaints, and cough or shortness of breath. The diagnoses given have been depression, post-traumatic stress disorder (PTSD), chronic fatigue, cognitive dysfunction, bronchitis, asthma, fibromyalgia, alcohol abuse, anxiety, and sexual discomfort. The initial response from the military and Veterans Administration was that there was no evidence of anything wrong physically, therefore, the problems were due to psychological conditions—a finding with which those with FM and CFS are very familiar.

According to a spokesman from the Veterans Administration in Washington, D.C., more than $100 million is being, has been or will be spent on research into Gulf War syndrome, also known as Persian Gulf War Syndrome (PGWS). Many of these research studies may have direct implications for those of us with fibromyalgia or chronic fatigue.

Dr. Daniel Clauw, a leading researcher in fibromyalgia and also Gulf War syndrome, stated in an article in the Summer 1997 issue of The Fibromyalgia Times, that "it's not really conceivable that a single toxin or an environmental exposure (it would have had to be an environmental exposure that everyone that went to the Gulf was exposed to) could have caused this illness because the rate of the development of the Gulf War illness had to do with two things."

He believes that these two factors are being a reservist and being female, and that these two factors have more to do with resulting health problems than a specific toxin. After every war soldiers come back with health problems. In the past these problems were called shell-shock or post-traumatic stress disorder or some other such name. It is a fact that when people are sent into a war, some of them come back changed. Until now, no one has done a comprehensive study of the effects of the stresses of war on individual service members.

However with the latest research that is being done on stress and the role that it plays in health, medicine is taking a new direction, a departure from the traditional Western concept of illness being either physical, with a clearly defined physiological base, or psychiatric, a solely mental problem. It is generally recognized now that it is nearly impossible to separate the two so that a new area of medicine is being opened and examined. Into this new area fall conditions like fibromyalgia, chronic fatigue, and now, Gulf War syndrome.

On February 5, 1998 the Committee on Veterans Affairs in the U.S. House of Representatives, heard a report from the Institute of Medicine's committee on the Department of Defense's handling of Gulf War syndrome. This committee, consisting of medical and public health experts, presented a number of recommendations. These recommendations came from two workshops on the topics and were made by leading scientific and clinical experts as well as the Department of Defense, the Veterans Administration, and other entities including representatives of veterans groups.

The first point made was that there were such a large variety of symptoms and systems involved that a term was necessary that would cover "difficult to diagnose" or "ill defined" conditions. It was decided to apply the term *medically unexplained symptom syndromes* to cover these health conditions. In some cases a specific diagnosis was made, specifically, of chronic fatigue syndrome, fibromyalgia or multiple chemical sensitivity, yet there is no clear objective finding for the cause of these conditions.

The committee felt that it must be recognized that there may never be a true cause or causes found, yet no one could deny that these veterans were indeed sick and many cases show significant impairment. Even with computers and sophisticated equipment, it may be nearly impossible to ever duplicate the exact conditions that occurred in the Gulf War.

The second point was that while many of these syndromes are often associated with depression and anxiety, they are not psychiatric disorders. Below are the Committee's recommendations. I include them here because of the impact they could very well have on fibromyalgia and the other associated conditions.

Medically Unexplained Symptom Syndromes

- The provider evaluating these patients must have access to the complete medical record including prior treatment.
- Rather than attempting to fit a treatment to a diagnosis, treatment should target specific symptoms or syndromes (e.g., pain, fatigue, depression.)
- A patient's functional impairments should be identified early to facilitate treatment.

Stress

- The initial CCEP (The Department of Defense's Comprehensive Clinical Evaluation Program) examination should include questions regarding traumatic event exposure. Any positive response should be followed up with a narrative inquiry.
- Stressors must be acknowledged as a legitimate but not necessarily sole cause of physical symptoms and conditions.
- Department of Defense (DoD) should provide special training and debriefing for those engaged in high-risk jobs during deployment, e.g., graves registration.
- DoD should provide risk or hazard communication to each about-to-be deployed soldier.
- Adequate time must be provided for provider/patient interaction.

Screening

- There should be increased screening for depression at the primary care level.
- Every physician should employ a simple, standardized screen for depression (e.g., BDI, Zung Scale, CES-D, IDD).
- Patients diagnosed with depression should be interviewed regarding traumatic exposure.
- Patients identified with any significant PTSD symptoms and/or a significant traumatic stressor should be referred to a qualified mental health professional for further evaluation and treatment.
- Every physician should employ a simple standardized screen for substance abuse (e.g., CAGE, brief MAST, T-ACE, TWEAK, AUDIT).
- Every patient who screens positive for substance abuse should be referred for further evaluation and treatment.
- DoD should explore feasibility of neurobehavioral testing at entry into the military for usefulness in measuring change in function.

Program Evaluation

- An evaluation should be conducted to examine: (1) the consistency of Phase I examinations across facilities; (2) the patterns of referral program from Phase I to Phase II; and (3) the adequacy of treatment provided to certain categories of patients where the potential for positive impact is great (e.g., depression).

Education

- DoD should explore ways to increase communication with the VA, particularly as it relates to the ongoing treatment of patients.
- DoD should examine the provider education materials and programs developed by the VA to determine if they might serve as models for DoD approaches.
- Education is needed to emphasize that conditions related to stress are not necessarily psychiatric conditions.

- Education should emphasize that depression is common and treat-
  able and that patients with depression can continue to function.
- CCEP (committee) information should be used to develop case
  studies that will help education providers learn about Persian Gulf
  health problems.
- DoD educational efforts should also address the concerns of
  Persian Gulf-deployed individuals and their families.

As mentioned earlier, Dr. Daniel Clauw, in addressing a group of individuals with fibromyalgia, and their families in Washington, D.C., in November,
1997, gave the two reasons why he believes there is a major difference in the
GWS from any previous veteran or military health problems. These are the
large number of reservists who were called up to duty in the Gulf and the fact
that a significant number of them were women. This is not to say that
reservists and women are any less soldiers than regular troops and men. But,
he contends, most reservists were unprepared to be thrust into a war zone
because it had been so long since there had been any actual combat involvement. These reservists were going about their normal lives while giving one
weekend a month with their annual two weeks in the summer for training
when they were suddenly called up to active duty and thrust into the con-
frontation with Saddam Hussein. Reservists usually don't have the intense
training and expectation of facing combat that regular troops do, so the
stress of actual combat is much higher on them. According to Clauw,
reservists were 1.5 to 2 times as likely to become ill as regular troops, while
women (soldiers, reservists or regular military) were 1.2 to 1.3 times as likely
to become ill (as male soldiers, reservists or regular military). The higher rate
of women falls in line with what is seen in many of these other conditions,
such as FM and CFS. Whatever the cause or causes are, female gender is a
factor. Because there is a high percentage of women in the majority of these
conditions, it would indicate there is a link that is tied into female hormones
or some other factor.

Some of these research studies have been completed, while others will
continue through the year 2000. Completed studies to date have indicated
that "the medical and psychiatric conditions identified have had a measur-
able impact on the functional activity and daily lives of Persian Gulf veterans"
but "may not be unique to the Persian Gulf War and may be similar to the
experiences of veterans of other wars." They also have found that while there
are some "minimal differences" seen between reservists and regular military
personnel, all military personnel were affected by the deployment.

A major concern for many veterans was the possible health impact on
their babies and children. A study which was reported in the New England
Journal of Medicine, in 1997, found that veterans who had been deployed
to the Gulf War theater were not more likely to have children with severe

birth defects than those Gulf War veterans who had not been deployed to the war theater. The study was based on live births, and did not include abortions or miscarriages. They also found the numbers of children born and the ratio of boys to girls both to be comparable to non-deployed military personnel, both male and female. Other studies are being done on reproductive outcomes.

One other completed study indicates a higher accidental death rate among Persian Gulf War veterans compared to those who served outside the theater of operations. Deaths of Gulf War veterans since returning from the war were found to be entirely due to external causes primarily to automobile accidents. According to Dr. Han K. Kang, Director of the Veterans Administration's Environmental Epidemiology Service, there was no "observed excess of suicides, homicides, or deaths from disease-related causes, and the rate of death from infections and parasitic diseases was slightly lower among Gulf War veterans than other veterans."

While the authors of the study had no explanation for the increased accidents, they felt it might be due to a "tendency for survivors of war to perceive risk differently and engage in more risk-taking behavior or that they may be at risk of post-traumatic stress disorder or depressive disorders." One aspect of the study did show a significant difference: among women veterans, there was an excess of deaths from all causes, including accidental deaths. There was also an increase in deaths from vehicle accidents, suicides, and homicides although it was not considered significant.

These studies examined the brain and nervous system; exposures to chemical weapons, depleted uranium, environmental toxins; immunology; Leishmaniasis; mortality; reproductive health; and symptoms and general health. Dr. Clauw believes that because the government's first reaction was to deny that troops were exposed to toxins or other risks, many veterans will continue to doubt anything the government says or does regarding the health of Gulf War veterans. Federal assistance, however, has been provided to many ill Gulf War veterans.

The VA has established a Persian Gulf Registry for those with health problems. By the end of February 1998, almost 227,557 veterans had gone to the VA for either routine health care or specifically for illness resulted from the Gulf War, while approximately 66,400 had taken advantage of the free complete physical exam being offered to all Gulf War Veterans whether they are ill or not. There is a toll-free information line to provide general information. Each VA hospital has one physician to coordinate exams for these veterans. Four VA medical centers—Washington, D.C., Houston, Los Angeles, and Birmingham, Alabama—have been established as Persian Gulf Referral Centers. A special program has also been established for certain spouses and children of Persian Gulf Registry participants and has been extended to December 31, 1998. This program provides payment for health exams by non-VA physicians in non-VA medical facilities. A second program records medical information on certain spouses and children provided by their private physician to the Persian Gulf Registry.

The VA is also providing disability payments to veterans who are chronically disabled. Any veteran whose claim had been denied previously should be reexamined. The expiration date for the development of any illness that might be linked to the Gulf War has been extended to the end of 2001. According to the VA, 90,665 Gulf War Vets are receiving 10 percent or more disability compensation. The VA is still reevaluating some claims which had been rejected earlier, either because of no definitive cause or link to the Gulf War or because they were beyond the usual two-year date of service. A fact sheet from the VA's web site indicated there were another 45,630 veterans who had served in the war theater and had service-connected disabilities but were not drawing a compensation check; of that figure, 37,253 served during the actual hostilities.

It may be that, although we would never wish any of our health problems on anyone else, the experience of the Gulf War Veterans may bring us much needed dollars for research.

Multiple Chemical Sensitivities

Multiple chemical sensitivity (MCS) is another new condition more commonly recognized in the last few years. It is defined as "An acquired disorder triggered by exposure to diverse chemicals at doses far below those documented to cause adverse effects in humans. Symptoms are recurrent, involve many organ systems, and are elicited by exposure to the offending chemical compound(s)." This definition was given in a study that compared CFS, FM, and MCS and was reported in the *Archives of Internal Medicine.* MCS mimics many of the symptoms of FM, CFS, and other DSS illnesses. The major symptom is a heightened sensitivity to different chemicals and other irritants. "Sick building syndrome" is a term that has been applied to MCS. Irritants can include air pollutants; fuels; building materials, like paints, carpeting, and cleaning solutions; and many scented products including perfumes and colognes. Many with MCS also react to food additives, foods they had previously been able to tolerate, drugs, and such elements as sunlight, loud noises, temperature extremes, and even touch.

Symptoms generally involve the neurological, immune, respiratory, and musculoskeletal systems and include chronic fatigue, aching joints and muscles, difficulty sleeping and concentrating, memory loss, migraines, and irritated eyes, nose, ears, throat and/or skin. Repeated exposure can increase the severity of the symptoms. According to an article in the *Fibromyalgia Frontiers* (Fall 1997), other disorders linked with MCS include asthma, chronic fatigue, fibromyalgia, migraine, Gulf War Syndrome, lupus, Akureyri disease, (benign myalgic) encephalomyelitis, cacosmia, disorders of porphyrin metabolism, epidemic neuromyasthenia, Icelandic Disease, mastocytosis, (postviral) neurasthenia, Royal Free (Hospital) Disease, silicone adjuvant disease, and toxic encephalopathy.

One of the more recent studies completed a profile of patients with chemical injury and sensitivity and indicated abnormalities involving the liver, nervous system (brain, plus limbic, peripheral, autonomic nervous systems), immune system, and porphyrin metabolism. Abnormal laboratory results indicate that this is not a psychologic disorder. This same study indicated that attention deficit and hyperactivity in children might be due to MCS. Treatment is presently tied to the reduction and environmental control of exposures.

Dr. Clauw has viewed MCS as a possible subset of FM but is reluctant to accept a diagnosis based on strictly environmental factors, because he believes it is much more complex that this.

Migraines/Tension Headaches

Tension and migraine headaches have been consistently noted in some people with fibromyalgia throughout the last 20 years of FM research. Many report chronic diffuse and bilateral headaches that are affected by emotional stress and are not caused by another condition. Tension headaches are considered to be on the lower end of the pain range, while the migraines are on the upper or more severe end. For those who suffer from either, the distinction may be one of semantics. The demographics for these headaches are similar to those found in the other conditions we have discussed. Headaches are considered one of the most common sources of pain in humans; more women suffer from headaches than men. When comparing tension, migraine, and cluster headaches—the three most common types of headaches—women have more tension and migraine headaches than men; men outnumber women in cluster headaches. Because cluster headaches do not fall under the umbrella of associated conditions, we will not be looking at them.

Usually headaches are fairly simple, the result of stress, eye strain, or fatigue and can be treated with minor doses of over-the-counter analgesics. But when tension and migraine headaches become more frequent, they have a definite impact upon both healthcare seeking and job performance. In one recent study, it was found that approximately 40 percent of the population experiences episodic, tension headaches but they were found most often in white, well-educated women in their thirties. As many as three-fourths of all premenopausal women experience headaches during their reproductive years, sometimes as often as once a month. Studies have shown a possible link with hormones, because these headaches are impacted by the menstrual cycle. Women in menopause may develop headaches when their body stops producing estrogen.

The study also showed racial differences for both men and women, but that regardless of race women were more likely to experience headaches than men. The study showed that about 8 percent of those with headaches lost work days (about 8.9), while 43.6 percent indicated they had significantly reduced-effectiveness days. With those who had the more chronic type tension headache, indicated by more than 15 headaches a month, 11.8 percent of the subjects reported a loss of 27.4 workdays each.

Not all pain in the head is due to tension headaches or migraines. Some can be attributed to sinus infections, toothaches, allergies, temporomandibular joint disorder, colds, or even caffeine withdrawal. Serious causes of head pain must be ruled out; these include meningitis, stroke, internal cranial bleeding, blood clot or tumor, high blood pressure, or temporal arteritis. A doctor should be consulted if there is any question of what is causing a headache.

Tension headaches are often caused or made worse by stress. They are usually located at the back of the head or it may feel as if there is a band pressing over the top of the head. Episodic headaches can begin in the afternoon and grow worse as the evening goes on. Contributing factors can be poor posture, eyestrain, stress or psychological problems. Whiplash is an injury that causes trauma to the cervical area of the spine, and it is not unusual for headaches to occur afterwards. (The cervical area of the spine and the upper part of the back and shoulders are common areas for FM pain.)

Treatment for tension headaches usually focuses on stress management and over-the-counter analgesics such as aspirin, acetaminophen, or ibuprofen. However, care must be taken so that a *rebound headache* doesn't complicate the issue. A rebound headache occurs when a person takes too much medicine in an attempt to control pain. If over-the- counter medications don't stop the headache or if either migraine or a combination tension-migraine headache is present, consult your doctor.

Although some tension headaches can become quite severe, migraine headaches can be extremely debilitating, involving numbness, sensations of prickling and tingling, nausea and vomiting, and a strong intolerance to light. They may be preceded by *auras*, which can appear up to an hour before the pain actually begins. Once present, migraines can last for several hours. Migraines occur when there are sudden changes in the blood vessels in the head. These changes are linked to serotonin, a neurotransmitter that acts as a vasoconstrictor. Sometimes inflammation is present. Treatment may include oral analgesics, nonsteroidal anti-inflammatory drugs, muscle relaxers, and/or vasoconstrictors.

Because headaches are often given as a symptom after some sort of stimuli, whether chemical, noise, or stress, one attempt at managing headaches is to watch for those stimuli and try to avoid or at least minimize the exposure. My sister has a severe response to various chemicals including perfumes and that response will often be a severe migraine. She is active in her church and answers the telephone for the children's day care center there.

However, her response to some smells has forced her out of the sanctuary for church services. The church set up a small television in a separate room where she is able to observe the services. Even with these precautions, she still requires powerful medications such as ergotamine plus an anti-nausea suppository once a headache has begun.

Dysmenorrhea/Vulvar/Vaginal Pain

Dysmenorrhea is the medical term for painful menstrual cramps. Often these begin within the first year or two of menstruation but may disappear by the time a woman reaches her twenties. The pains are crampy, spasmodic episodes of pain in the lower abdomen that may spread to the lower back, hips, and even thighs. They are often accompanied by diarrhea or constipation and, in the extreme, may include vomiting and fainting.

If cramps are very severe, tests should be done to rule out such factors as endometriosis, adenomyosis, pelvic inflammatory disease (PID), an improperly placed intrauterine device (IUD), ovarian cysts, or uterine fibroids.

In the absence of disease, generally treatment includes aspirin or acetaminophen and the application of a heating pad to the abdomen. If there is no obvious abnormality in the reproductive organs, the cause of dysmenorrhea is generally attributed to an excess of prostaglandins, normally occurring fatty acids, that cause uterine contractions and stimulate the intestines. Aspirin is a mild antiprostaglandin. If aspirin or acetaminophen does not ease the pain, something stronger may be used, such as nonsteroidal anti-inflammatory drugs like ibuprofen, naproxen, or mefenamic acid, either available over-the-counter or in stronger prescription-strength doses.

Other non-drug treatment includes stress management, biofeedback, or relaxation exercises. A position that I instinctively used years ago is also recommended: lying with knees elevated, or on the side curled up in the fetal position. I can also vouch for the benefits of cups of hot tea.

Dr. Clauw believes there is a temporal link involved with some of these conditions. In children the first symptoms are musculoskeletal "growing pains" and then painful menstrual cramps in teenage girls. These women often go on to develop FM as they are exposed to more stressors of various sorts in their lives, until, by their 40s and 50s they have full-blown fibromyalgia.

Another area of pain for women occurs in the vulva, including the vaginal opening and the urethra. *Vulvodynia* is the medical term for this pain. For years women were told this pain was (here we go again!) "all in their heads," yet it could and often did have a major impact on their intimate relationships with either husbands or lovers. Often they were referred to psychiatrists and told they were frigid or afraid of sex. According to the *Harvard Guide to Women's Health*, this condition may have actually been identified as early as the 1880s and again in the 1930s.

The *Guide* also states that it is probably a complex condition that may be caused by: "abnormally high levels of calcium oxalate crystals in the blood and urine, chronic infections with the herpes simplex virus, an autoimmune response to yeast or human papilloma virus (HPV), or *disturbances of chemicals in the brain called neurotransmitters.*" (My emphasis added.)

Vulvodynia is likely to be present in women with a chronic pain condition or women with interstitial cystitis. Symptoms include burning, painful, stinging sensations in the vulvar region, which may also be dry, irritated, or raw. Itching may or may not be present. If symptoms are severe, sexual relations can be so painful that the woman avoids them. The pain can also make sitting, walking, and wearing tight pants uncomfortable. Depression is often seen in women with this condition but as in related DSS illnesses, the question is which came first, the depression or the painful condition?

A diagnosis is reached by ruling out other possible physical causes. For these women, pelvic exams and pap smears can be very uncomfortable because even the touch of a cotton swab is painful. A urine sample is tested for the presence of calcium oxalate. If the physician wishes, a computed tomography (CT) scan or magnetic resonance image (MRI) of the lower vertebrae rules out tumors or cysts.

Treatment may include low doses of antidepressants such as Elavil. If there is a physical cause, treatment should be directed to reversing it. Sometimes this may involve surgery, or local injections of interferon alpha 2b, which may cause local pain or flulike symptoms. If high levels of calcium oxalate are the problem, diet restrictions may be necessary to rule out foods that contain oxalic acid (oxalate), or medication such as calcium citrate may be given to reduce the calcium oxalate levels.

Thyroid Disease

Thyroid disorders have long been one of the alternative diagnoses when fibromyalgia symptoms appear, basically because they share many common symptoms. Women are more likely to have thyroid disorder than men. Although hyperthyroidism and hypothyroidism are on opposite sides of dysfunction, some of the symptoms from each have been found in FM. In hyperthyroidism, increased levels of thyroid hormone speed up the body's metabolic rate and produce weight loss, heart palpitations, mood swings, muscle weakness, and diarrhea. In hypothyroidism, a deficiency of the hormone causes weight gain, fatigue, intolerance of cold temperatures, muscle cramps, and ultimately life-threatening coma.

Doctors from the Departments of Physical Medicine & Rehabilitation at Riverside Methodist and the Ohio State University (both in Columbus) conducted a study on the relationship between FM and thyroid functions. They found that women with FM reported a history of thyroid disorders more frequently than a group of controls. They reported more symptoms than are normally associated with either of these disorders more often than the controls, but these symptoms were not specific. More recent studies found that some individuals with FM have a slightly lower ability to process thyroid hormones (T3 uptake evaluation), but the significance has not yet been found.

Another study published in the *Scandinavian Journal of Primary Health Care* (June 1996) tested the association between chronic widespread musculoskeletal complaints and thyroid autoimmunity. They found a significantly higher prevalence of thyroid microsomal antibodies in those with chronic widespread pain, although the thyroid function tests did not differ between the two groups. The study concluded that there may be a subset of patients in which there is thyroid autoimmunity rather than thyroid dysfunction (an important distinction). They believe fibromyalgia could play a role in this situation.

Temporomandibular Joint Dysfunction

Temporomandibular joint dysfunction (TMD or TMJ) is a myofascial disorder with the following symptoms: pain in the forehead or temple with a migraine or sinus-type headache; jaw pain; pain behind the eyes with, perhaps, some sensitivity to light; a dull or diffuse earache; ringing in the ears; and dizziness. The jaw may have limited or painful movement and there may be clicking or popping in the jaw joints. There may also be pain in the cheek muscles, which may or may not be related to clenching the jaw or grinding the teeth together, particularly at night. There may be difficulty in swallowing and pain and/or stiffness in the neck and upper back and shoulders. More women than men have TMD. It is usually diagnosed in the 20s and 30s. If TMD is present with fibromyalgia, this regional pain problem can be compounded.

A report in the *Journal of Rheumatology* found that only 18.4 percent of the subjects with TMD also had FM, while those patients who met the ACR criteria for FM had a 75 percent rate of also having TMD. In other words, only a small percentage of individuals whose primary diagnosis is TMD also have fibromyalgia, but of those individuals who have FM, there is a high percentage who also have TMD. Those with FM who also had TMD had higher levels of severity in pain, fatigue, sleep disorders, as well as functional disability, work difficulty, and general dissatisfaction with health.

Causes of TMD vary and range from trauma, to loss of teeth or poor fitting dentures, to degenerative joint diseases. Treatment includes analgesics such as aspirin or acetaminophen; muscle relaxants; topical sprays to relieve or mask the pain; or nonsteroidal anti-inflammatory drugs if there is inflammation of the joint. A plastic mouth guard may be used to prevent clenching the jaws. Warm compresses may help ease the pain. Stress management or relaxation techniques may ease the symptoms if stress is a factor.

Mitral Valve Prolapse (MVP)

Mitral valve prolapse occurs when one of the leaflets or valves that allow blood to flow between the upper and lower chambers on the left side of the heart becomes damaged and no longer closes properly. The valves may be scarred from a congenital defect or after an infection like rheumatic fever, a disease which follows a strep infection, usually during childhood. Rheumatic fever is less common now and is usually treated quickly, thus preventing valve damage. In other cases, there may not be a clear cause, yet the valves may bulge and even allow some blood to flow back into the chamber.

Although sources disagree on the exact percentages, all agree that MVP occurs most often in women with rates ranging from 5 to 20 percent depending on which source you reference. Sometimes there are no symptoms, and unless a doctor finds the distinctive click-murmur during an examination, these women may never be aware of the condition. In other women, however, there may be chest pain, shortness of breath, dizziness or lightheadedness, fatigue, anxiety, or panic attacks. If a doctor does not pick up the sounds and make the diagnosis of mitral valve prolapse, the woman may be sent to a psychiatrist.

If mitral valve prolapse does present problems, treatment will usually include a round of antibiotics whenever any dental work or certain other invasive exams or surgeries are to be done. This is to prevent an infection from spreading to the heart, a situation that can be very serious. At this point, the link between MVP and FM is not clear, but it is commonly found in individuals with FM and several of the other conditions listed in this chapter. One theory, offered by Dr. Devin Starlanyl and Mary Ellen Copeland, M.S., M.A., in their book *Fibromyalgia & Chronic Myofascial Pain Syndrome* is that the mitral valve itself is made of connective tissue. Since FM affects muscles, tendons, ligaments, and the connective tissue (called fascia) which surrounds them, it is possible that FM somehow impacts the connective tissue of the mitral valve causing it to "lose its elasticity" as Dr. Starlanyl states, "and it no longer functions normally."

Regional Fibromyalgia/Myofascial Pain Syndrome

Myofascial pain syndrome (MFS) occurs when trauma injures the muscles or tendons and ligaments in a localized area such as the cervical area of the spine, the shoulder, or hip. Probably the best known example is lower back pain not produced by changes in the vertebrae of the spinal cord. Myofascial pain is characterized by trigger points that produce referred pain, pain that is felt at a location different from where pressure is applied or where the true problem lies.

One of the leading theories regarding fibromyalgia is that it may begin as a localized myofascial that isn't resolved with treatment. The spillover of

pain-message neurotransmitters to nearby tissues makes tissues extra sensitive to stimuli.

Cervical Spinal Cord Compression

Dr. Michael J. Rosner, of the University of Alabama, recently reported on the
presence of *cervical spinal cord compression* in patients with fibromyalgia
and chronic fatigue syndrome. He and his colleagues found that two patients
with cervical stenosis (compression) had improvement in their chronic fatigue
symptoms when corrective surgery was performed. He decided to test the
hypothesis that FM is caused when the cervical spinal canal is compressed,
thereby affecting the flow of cerebrospinal fluid to the brain. Of 48 patients
with either chronic fatigue or fibromyalgia, all had possible *myelopathy* or
disease of the spinal cord.

Other findings include hyperreflexia of the neck, a neurological condition
marked by increased reflex actions, Babinski and other motor neuron findings. The motor neurons are nerve cells which carry nerve signals from the
brain or spinal cord to muscle or gland tissue. Babinski's reflex or sign refers
to the extension of the big toe upward and fanning the other toes when the
sole of the foot is stroked. This is a normal sign in a newborn but in adults
may indicate a brain injury or, as postulated here, a problem in the transmission of nerve signals along the spinal cord. Seventy to 75 percent also had
sensory and/or motor dysfunction in the upper and lower extremities, abnormalities of gait, and other defects, including narrowing of the spinal canal in
the mid-cervical region. There was improvement in neurologic symptoms and
signs of normal or nearly normal functioning after surgery decompressed the
cervical spinal canal with 60-65 percent of the patients. Levels of change were
checked by means of a questionnaire at two points after the surgery. In the
final outcome, nearly all of the patients had improvement in some areas, and
more than 50 percent felt overall improvement. Between 10 and 15 percent
felt they were worse in some areas, while another 10 to 20 percent reported
no change in some symptoms but improvement in others.

The surgical procedures, performed by neurosurgeons, varied depending on the specific findings for each patient but the object was to increase the
spinal cord diameter to improve the flow of cerebrospinal fluid. Dr. Rosner's
findings were supported in another study, presented at the 1997 meeting of
the American College of Rheumatology held in Washington, D.C. That study
compared individuals with FM only, with Chiari's disease only or FM and
Chiari's disease. This study concluded that "Chiari tends to be associated
with FM in patients in a rheumatology clinic, although it generally does not
elevate levels of CSF substance P, pain, or fatigue." The study recommends
that "all FM patients should be examined for signs of bulbar/spinal cord
compression and those with evidence of compression should be referred for
MRI" because "a small number of the FM patients with Chiari may require

and benefit from surgery." Chiari Malformation or syndrome refers to the space in the back of the cervical spine which is smaller than it should be and through which the spinal cord runs. This can be a congenital condition, present since birth, or arise from some trauma, such as a whiplash injury.

There is now an FMS spinal cord compression surveillance research study underway, run by Dr. Robert Bennett, Oregon Health Sciences University; Dr. I. Jon Russell, University of Texas Health Science Center at San Antonio; and Dr. Daniel Clauw, Georgetown University, Washington, D.C.; and coordinated by Dr. Rosner. This research is funded by the National Fibromyalgia Research Association of Salem, Oregon. It is possible there is a subset of individuals with FM or chronic fatigue syndrome who have cervical spinal cord compression.

Psychological Abnormalities

In 1987, when I first began to research fibromyalgia, a member of the local chapter of the Arthritis Foundation shared a quote with me showing that many of the medical community had long regarded FM sufferers as "bitchy women whose symptoms were all in their head." For years many of us have had to fight that image. And, finally, for the most part, we—or at least the researchers—are proving them wrong. There are still some doctors who insist there is no physical basis for FM and that it must be psychological. It's hard to see how they can be so blind when so many studies are proving otherwise. It is also hard to understand how they can ignore the mind/body link, which has always been a part of many medical practices and beliefs, even though Western doctors have ignored it until the last few years.

Regardless of some doctors' attitudes, the majority of physicians and almost every researcher of fibromyalgia and chronic fatigue knows that these syndromes have a physiological basis, and this book presents those scientific findings. But there is a role that psychology and psychological factors play in FM and chronic fatigue syndrome (CFS). Most who have fought for legitimate recognition of their symptoms and problems as physical are rather "gun shy" when the word "stress" is mentioned. But as many studies show, stress can and does have a physiological impact.

The mind and body are inexorably linked and no pain, no pleasure, indeed no aspect of life, can occur without an interaction between the two. Human beings are not just physical beings. Humans think, react, and change based on the experiences of life, beginning with emotions learned from parents, siblings, other relatives, friends, and peers. Each culture, each ethnic group, and ultimately each family contributes to how an individual will react to life and the normal stressors, good or bad, that come with living. Yet, in the end, each individual is unique, also bringing his or her own personality and beliefs to how he or she faces life.

To understand the psychological aspects of fibromyalgia, we need to look at the psychological tests that were administered when it was still suspected that FM was a mental health problem. The primary test, the *Minnesota Multiphasic Personality Inventory* (MMPI) had a number of questions that dealt with sleep, fatigue, and musculoskeletal pain. Because these are the basic factors in fibromyalgia, most individuals would indicate that they did not get restorative sleep, they were tired all of the time, and that they hurt "all over." It took several studies, led by Dr. Frederick Wolfe, Dr. Muhammad Yunus, and Dr. Hugh Smythe to prove that these tests were inappropriate for those with FM or CFS.

For example, the following five statements were included in the questionnaires and it was determined that they are really related to the physical disease and were found consistently in the profiles of those with rheumatoid arthritis as well as those with FM.

1. I wake up fresh and rested most mornings.
2. I am about as able to work as I ever was.
3. I seldom worry about my health.
4. I am in just as good physical health as most of my friends.
5. I have little or no pain.

Almost anyone with fibromyalgia would respond to these statements negatively. Response patterns to the five disease-related items are associated with the severity of pain, determined by health assessment questionnaire (HAQ) scores, and grip-strength measures, as well as the presence of disease. Since that time, new tests more appropriate for anyone with a chronic pain condition have been developed.

Some studies have shown that there is a relationship between fibromyalgia and *major affective* (mood) disorder as well as a familial predisposition to both depression and fibromyalgia. The general consensus is that fibromyalgia and its associated *psychopathology* (mental illness) are caused by a common but unknown abnormality. This ties in with Dr. Yunus' theory of dysregulation spectrum syndrome (DSS) and Dr. Clauw's belief that there is a shared cause or causes for fibromyalgia, chronic fatigue syndrome, and many of the other conditions listed in this chapter.

Numerous studies since have proved that while a group of those with FM have depression or anxiety, the percentage is no higher than those found in the general public or even in other chronic health conditions. Many recent studies show that often the psychological distress is directly related to pain severity. In one study by Celiker et al., it was determined that pain severity was found to be "correlated with trait anxiety inventory (distress) scores." The authors felt that the "difference between state and trait anxiety inventory reflects that current anxiety is not secondary to pain but trait anxiety is possibly causally related to pain."

Another study reported in January 1997 compared tender points and depressive and functional symptoms between patients with fibromyalgia and those with major depression. The patients with FM had markedly more tender points than the depressed patients. The report concluded that an "increased sensitivity to pressure pain *clearly distinguishes fibromyalgia from depression even if there is an overlap of other symptoms.*" (My emphasis.) Individuals with depression showed some clinical pain and there were both depressive and functional symptoms in both groups but there was a clear difference based on the tender points.

Drs. Aaron, Bradley, and others found that "psychiatric disorders were not intrinsically related to the FM syndrome. Instead, multiple lifetime psychiatric diagnoses may contribute to the decision to seek medical care for FM in tertiary (specialized) health care settings."

Several studies looked at FM patients in light of the type of symptom onset, trauma, illness, or emotional stress. One reported in *Pain* (December 1996) found that post-traumatic FS patients reported "significantly higher degrees of pain, disability, life interference, and affective distress as well as lower levels of activity than did the idiopathic FS patients." They concluded with the suggestion that "post-traumatic onset is associated with high level of difficulties in adaptation to chronic FS symptoms and that FS patients are a heterogeneous group of patients." While there have been other studies which have found no major differences in individuals with FM and different precipitators, this reinforces the idea that there are subgroups within the diagnosis of FM and the means of onset may be one way to separate those groups. One finding from the study under discussion was that the differences between the groups were not related to compensation status, concluding that "two types of FM onset cannot be clearly a function of secondary gain."

One study inquired into the prevalence of FM in relation to nonarticular tenderness, symptoms, quality of life, and functional impairment among post-traumatic stress disorder patients. PTSD patients were compared with normal controls and tested for FM. There was no one in the control group who met the criteria for FM but within the post-traumatic group there was a 21 percent incidence of FM. The PTSD group also had more muscle tenderness than the control group. Those who met the criteria for FM had even greater muscle tenderness. Those with FM also reported more pain, lower quality of life, more psychological distress, and more functional impairment than the PTSD without FM. The report urged checking PTSD individuals for FM. This study emphasizes the link between psychological stress and pain syndromes. Note that this says there is a *link*, not that the pain is psychological.

One of the primary questions asked is whether depression is the cause of the pain, or if pain and chronic ill health have depressed the individual. A study published in the *Scandinavian Journal of Rheumatology*, 1996, by Carol Burkhardt and A. Bjelle, and reviewed by Dr. Robert Bennett in the *Journal of Musculoskeletal Pain* showed that "FM patients perceived a poorer level of control over their symptoms than those patients with RA(rheumatoid arthri-

---

Conclusions from Prior Studies of the Psychological Associations with Fibromyalgia and Suggestions for Future Studies

- The majority (65 to 80 percent) of people with fibromyalgia do not have an active psychiatric disorder. A subgroup of people may have significant active psychopathology. This group should be better delineated and evaluated.
- There may be a greater prevalence of depression in people with fibromyalgia compared with rheumatoid arthritis, although some studies have not found such differences. Most studies find a greater prevalence of symptoms of depression in those with rheumatoid arthritis, fibromyalgia, or any chronic medical illness than in people with other medical problems.
- There may also be a greater lifetime history and family history of depression in people with fibromyalgia compared to rheumatoid arthritis patients and controls. These may indicate some psychological link that should be investigated.
- Psychological tests such as the MMPI and even a structured interview technique such as the diagnostic interview schedule (DIS) do not adequately work for chronic pain and associated medical conditions. Such tests must be modified for use in medical patients and then validated in various chronic pain disorders. Previous reports of psychopathology in people with fibromyalgia may have affected disease activity reporting.
- Future studies should select psychiatric tests that are less affected by chronic pain. The effect of medications on symptoms of pain should be distinguished from their effect on mood and psyche. Antidepressant doses could then be better evaluated regarding their effect on fibromyalgia. The psychiatric state of a person should be known, either by prospective studies or by structured historical interview data.

Courtesy of Dr. Don Goldenberg. Reprinted with permission from *Rheumatic Disease Clinics of North America.*

---

tis) or SLE (lupus). Compared to the other two diseases, FM patients had inferior quality of life, high levels of depression and anxiety, and impaired coping strategies on the Coping Strategies questionnaire." In his comments Dr. Bennett stated "This study would suggest that this negative impact is in part related to poorer coping mechanisms and a feeling of helplessness. It would be interesting to know *whether society's and the medical profession's often pejorative perception of the fibromyalgia syndrome is in part responsible for these attitudes. The chicken and egg dilemma is probably relevant to inter-*

pretation of these findings." (My emphasis.) I would like to know the answer to this myself.

Pain is both a sensory and an emotional experience. Although the sensory side may have a strong physical aspect, according to recent research, even that depends upon the emotional aspects involved. Does this mean that if I cry when I'm in pain, it's only emotional? No. But neither can we say it is strictly physical.

At the ethnic and cultural level, reaction to pain can range from a silent, stoic bearing to the most vocal expressions possible including cries, screams or curses. Few people think anything about it if someone who is injured cries or reacts with tears. But when pain becomes chronic, the same reactions are no longer acceptable. Therefore, new pain behavior is developed. Some of that pain behavior may not be healthy, leaving us angry, depressed, or anxious. We need to learn proper, healthy ways of coping with the limitations of our chronic illness and recognize our need for medical help with the results of our anger.

Other Conditions

There are some other conditions in which fibromyalgia may occur, but the link is more uncertain. A number of women have reported decreased health because of silicone breast implants. Numerous studies have been conducted yet no conclusive explanations can be made for all of the symptoms reported. One of the most common complaints has been the development of rheumatologic disorders. One study published in *Neurology* (1996), looked at the case histories of 131 women with the breast implants. Symptoms were as follows: fatigue, 82 percent; memory loss and other cognitive impairment, 76 percent; generalized myalgias, 66 percent. The tests conducted on the women showed that 66 percent had normal neurological exams and indicated that the symptoms were all "mild and usually subjective." Diagnoses were not standardized, nor according to the report, based on any "standards accepted by the neurologic community." The report stated that 9 cases could be fibromyalgia, 16 depression, 7 radiculopathy, 4 anxiety disorders, 4 multiple sclerosis, 1 multifocal motor neuropathy, 1 carpal tunnel syndrome, 1 dermatomyositis, and 3 others psychiatric disorders.

Another study reported in *Annual of Plastic Surgery* (July 1997), found that 18 out of 100 consecutive women who requested that silicone breast implants be removed developed health problems. These 18 were referred to rheumatologists after a diagnosis of either an autoimmune or rheumatic disease. According to the report, there were two patients with systemic lupus, two with rheumatoid arthritis, one with multiple sclerosis, and one with Raynaud's disease, while there were twelve with rheumatic disease—ten with FM and two with inflammatory arthritis. All symptoms developed after the breast implants.

As with research into fibromyalgia and chronic fatigue, not all studies agree. According to "Breast Implant Update" in *Harvard Women's Health Watch* (September 1997), most of the information up to this time had come "from observational studies." It concluded that "As the FDA notes, the jury is still out. It may be at least another decade before the results of controlled trials, which should provide more conclusive evidence, are in."

Two recent studies looked into the incidence of fibromyalgia in patients with *hepatitis C virus infection* (HCV). One published in *British Journal of Rheumatology* (September 1997), examined the prevalence of HCV in 112 FM patients with matched rheumatoid arthritis (RA) patients and then looked for FM in 58 patients diagnosed with chronic hepatitis due to HCV, compared with matched surgery clinic patients. HCV antibodies were found in 15.2 percent of the FM patients and in 5.3 percent of the RA controls. Fifty-three percent of the HCV had diffuse musculoskeletal pain, with 10 percent meeting the diagnosis of FM. In the control group 22 percent had diffuse musculoskeletal pain, and one female patient (1.7 percent) met the criteria for FM. The report concluded "There were no statistical differences in autoimmune markers between patients with and without FM. These data suggest that there exists an association between FM and active HCV infection in some of our patients. FM is not associated with liver damage or autoimmune markers in these patients. HCV infection should be considered in FM patients even though ALT (alanine aminotransferase) elevations were absent."

A second study reported in the *Archives of Internal Medicine* (November 1997), found that "a high prevalence of FM was observed in patients infected with HCV, especially women. Recognizing FM in patients with HCV will prevent misinterpretations of FM symptoms as part of the liver disease and will enable the physician to reassure the patient about these symptoms and to alleviate them."

These are good examples of cases when symptoms are not all due to fibromyalgia, and the doctor must look further to find and treat another condition. The opposite also applies.

Lyme disease is the most common tick-borne disease in parts of the United States. Although research continues since the National Institute of Allergies and Infectious Diseases first isolated the spirochetal (spiral-shaped) bacterium, *Borrelia burgdorferi*, as the cause, the course of the disease is still unclear. Some patients have the disease and recover, while others go on to develop persistent arthritis, nervous system problems, and even heart problems. There have been cases of fibromyalgia as well. Much research must be conducted on all of the variables of Lyme disease, and perhaps some of these variables will determine how certain individuals go on to develop FM.

A few studies have examined the possible link between individuals infected with the human immunodeficiency virus (HIV) and subjective rheumatic complaints. Dr. Gregory Gardner et al, at the University of Washington, Seattle, conducted surveys of HIV-positive subjects for arthralgias, myalgias, non-restorative sleep, and fatigue. They found that almost half of their patients reported the last three symptoms, while people who were at high risk for HIV infection (but tested negatively) did not report any of the symptoms. Dr. Gardner believes this means that the symptoms are either disease-or treatment-associated but he was not able to determine exactly how. The report states "The only finding of significance was the association between a history of current or past psychiatric diagnosis with non-restorative sleep." It also stated that the small size of the group (53 patients) might have led to an error in the study. In reviewing his study and others, Dr. Gardner found that only a minority (8 percent) of their patients actually met the criteria for FM but that "other studies have placed the prevalence of FMS in HIV-infected people between 10 and 29 percent compared to the general population figures of 3.4 percent for women and .5 percent for men." He concluded that his group wishes to study the possibility that "mood disturbance may contribute to the above symptoms in our patients and serve as a possible site of intervention for improving their quality of life." This is not to say that people with FM will become HIV-positive. It is saying that something in the treatment or progress of HIV-positive patients causes the FM-like symptoms to appear. Research into this area may give us one more piece of the puzzle regarding FM.

Summary

As you can see from all of these conditions, there is a wide spectrum of possible health problems that may or may not tie in to fibromyalgia. Not everyone is going to experience every one of these conditions or even a majority of them. Whether or not they share a common cause or set of causes, as Dr. Yunus and others speculate, we hope that you are now more aware of related conditions you might encounter. I have to confess that even after more than ten years of having a firm diagnosis of FM, and being aware that some of these conditions occurred, I learned more researching this chapter than I expected. I can also recognize some of the temporal links that Dr. Clauw discussed. I had growing pains as a child, painful menstrual cramps as a teenager and young adult, and increased problems with my FM symptoms as I grew older. Now at age 50, I experience significant problems with my FM. I know more than I did but I still have a lot of questions I want answered. Let's hope that some of that research the government is paying for in the Gulf War Syndrome or Medically Unexplained Symptom Syndrome will pay off for those of us already diagnosed with fibromyalgia.

FIVE

Fibromyalgia in Children

Although Dr. Muhammad Yunus reported the existence of fibromyalgia in children in 1985, there have been limited studies done in that area until the last few years. It is significant to note that children with fibromyalgia exhibit many of the same symptoms as adults. Widespread pain, fatigue, non-restful sleep, stiffness, numbness, feeling of swelling in fingers and hands, and headaches all occur in children. They also have irritable bowel and frequent urination. Some also meet the criteria for chronic fatigue syndrome, with sore throat and swollen lymph glands without evidence of an infection. It is not unusual for these children to be depressed and/or anxious. FM has been found in children as young as five and although the figures have been too small to give a total picture, more girls than boys seem to be affected.

Of the studies that have been done, it appears that the onset also copies that of adults, appearing after some sort of stressor such as accident or viral illness. There has been particular attention paid to the families of these children, their environment, and pain history. A number of recent studies show a familial link, with several family members likely to have FM. Because of a number of studies linking fibromyalgia in adults with some sort of abuse—sexual, physical, or emotional—during their childhood, studies on children have looked at the social context of pain as well as traumatic events during the child's life.

Of two studies presented at the American College of Rheumatology's 1997 conference, the one conducted by Duke University Medical Center, Durham, North Carolina, concluded that there was a definite link between family environment and parental pain history. The authors of the study could not determine if there was a cause-and-effect relationship, or if the presence of illness affected the family.

The other study, done in Brazil by Dr. Roizenblatt and colleagues, found increased levels of anxiety in children with either fibromyalgia or juvenile rheumatoid arthritis but reported that the children with fibromyalgia had more depression. Again, they were unable to determine the cause-and-effect relationship. Mothers of all children with chronic pain showed more anxiety than did controls, but mothers of children with FM had higher levels of depression. The researchers could not determine the cause-and-effect relationship in this situation either. A study in Finland corroborated the findings of increased diagnosis of depression in children with fibromyalgia and recommended that this depression be recognized and treated along with the FM.

A number of studies over the last few years demonstrate that when an individual experiences a higher level of traumatic events or stressors, that person is more likely to develop a physical illness. This has been the subject of some FM studies as well. Although it was first assumed that abuse in childhood—particularly sexual abuse—might be a prediction for someone to develop FM as an adult, this has not been proven. In fact, a study conducted by members of the Department of Psychiatry and Behavioral Sciences at the University of Washington in Seattle looked into the psychosocial factors for individuals with FM compared to those with rheumatoid arthritis. The study found a higher prevalence of all forms of victimization in both the childhood and adult life of the individual; experiences of physical assault in adulthood had the greatest correlation with unexplained pain.

With that thought in mind, a look into any traumatic events that have involved a child with FM will help make a diagnosis and formulate a treatment plan.

Obtaining a correct diagnosis is as important with children and adolescents as it is with adults. One study indicates that 7 percent of all pediatric office visits are because of musculoskeletal complaints. Some pediatricians may be reluctant to apply the American College of Rheumatology's Diagnostic Criteria for FM to children, particularly with regard to the tender points, believing the criteria has been developed for adults. One recent study found that instead of the 4 kg of pressure (measured by dolorimeter) used for adults, a lesser force of only 3 kg is sufficient to determine the presence of tender points in children. This finding must be confirmed with other studies. Many children have been incorrectly diagnosed with chronic Lyme arthritis, particularly in areas where Lyme disease is often found. Children who do have Lyme disease should also be examined for tender points and other symptoms of fibromyalgia, because the antibiotic treatments for Lyme disease have no impact on FM. Hypermobility and fibromyalgia were found in children in a study conducted by the Louisiana State University Medical Center in New Orleans and done in Beer-Sheva, Israel. Joint hypermobility means loose joints, which allows the muscles to be stretched beyond their normal range of motion. Of 338 children, 13 percent had joint hypermobility and 6 percent had FM. In those with FM, 81 percent had joint hypermobility and 40 percent of those with joint hypermobility had FM. No one knows yet how these two conditions may be linked.

Other incorrect diagnoses include juvenile rheumatoid arthritis, low back pain, hysteria, reflex sympathetic dystrophy syndrome, anterior knee pain, tendinitis, hypothyroidism, and inflammatory arthritis. It is also common for them to receive a diagnosis of "growing pains" or psychological problems. Sounds similar to what many adults have heard.

A study done in Italy and published in 1996 found that 1.2 percent of the students who answered a questionnaire sent to children in third grade to high school met the criteria for fibromyalgia. Another study, published in *Pain* (1997), conducted a follow-up study of children who had initially reported pain at least once a week, whether regional or widespread. Of the original 1,756 third- and fifth-graders, 1,626 took part in the follow-up. There were 564 children who had originally reported regional pain. Two hundred seventy of them (52.4 percent) indicated they still had pain. Widespread pain, meeting the criteria for FM, was reported by 132 children or 7.5 percent initially, and was still present in 35 of them or 29.7 percent in the follow-up 1 year later. The report also found that disability was more severe in children with widespread pain and indicated that almost one-half of the group who had reported any pain were still experiencing pain at the 1 year follow-up date.

Dr. Daniel Clauw, Georgetown University, Washington, D.C., believes there is a temporal link to fibromyalgia and that its earliest manifestation is in childhood "growing pains." When girls began their menstrual cycles, they also developed painful monthly cramps and signs of irritable bowel. Many women have reported that they had these growing pains as children and they can also remember being tired more often.

It is important to remember that although many of the studies are looking at trauma, stress, and family environment, no one is saying that fibromyalgia is to be blamed on parents or the family. There is no room for blame. The primary concern should be to achieve as much control over the symptoms as possible and to encourage a child to live as full and complete a life as he or she wants and is able despite the limitations of FM.

Every child is different. Some have fairly minor involvement with FM and others have more dysfunction. Treatment outcomes in children with FM have not been studied enough to make definitive statements; however, many children do improve. Quite often, in adults, if the presence of FM is diagnosed soon after it develops, there may be significant improvement. This would seem to apply to children as well. The usual treatment for adults is tricyclic antidepressants to improve sleep, aerobic exercise to improve cardiovascular fitness, and cognitive behavior training to deal with stress. This is also the standard for children, although adjustments have to be made. The dosage of medicine may have to be lowered for smaller body weight and different metabolism. Simple analgesics can help with pain relief. Stretching and moderate exercise along with proper sleep are all vital. It may be appropriate to include massage, posture retraining, and physical therapy as needed. As with an adult, treatment must be structured for the individual.

Modulating factors are generally the same for children as they are for adults: stress, cold, damp weather, changing weather, and either overactivity or inactivity. The child should not be allowed or encouraged to quit all physical activities. He or she should remain as active as possible and remain involved in as much as he or she can handle.

Adults feel a loss of control when FM invades their lives. Children, who have so very little control of their lives, must have an opportunity to exert some control in the case of FM. While they need to keep as active as possible, parents must allow them to choose when to participate in physical activities and when to rest. One researcher on FM advised not giving children a diagnosis at all, so they did not have some sort of built-in image of disability to live up to. I know that sometimes children can be lazy, they may work to get out of doing something, and they can also be a bit manipulative. Hey, guess what? So are some adults.

It is up to parents to recognize when their child is "trying something," but it is also up to them to respect their child enough to know when they are in pain and when they are tired. Children don't want to be different. They don't want to stand out for the other kids to pick on or to laugh at them. This is where doctor, parents, and teachers must all work together to help the child do as much as he or she can without overdoing and then paying the price physically.

Once a diagnosis has been made and a multidisciplinary plan of treatment has been established, parents need to sit down with their child and set up guidelines on how they will deal with school, play activities, teachers, and schoolmates. Proper coping skills should be taught including stress management, relaxation, biofeedback, and guided imagery where appropriate.

Because school is such a major part of a child's life, this is where most adjustments may have to be made. Schools often attempt to be as regimented as possible to maintain control and accomplish their primary objective of educating young minds. But this can present a problem when a child is different or has limitations. It is even harder when that child looks normal and healthy. Take the opportunity to speak with the teacher and the principal, if necessary, to explain about the fibromyalgia and the nature of this syndrome. It would also help to find a way to explain FM to the children with which your child associates.

If possible, make arrangements for rest periods and extra minutes between classes, in the higher grades. A backpack or a small cart on wheels is a better way to carry books than in one arm. If possible have two sets of books, one at school and one at home so he or she won't have to carry them to and from school. Remember that schools must make accommodations for children with disabilities. Occupational therapists will be able to help evaluate both the school and home environments to help save energy and find the best way for the child to continue with as much of their daily routine as possible. If you don't have access to an occupational therapist, contact the nearest Arthritis Foundation Chapter or Branch and obtain some of their literature on fibromyalgia or arthritis. Although FM does not cripple or damage joints, many of the principles regarding energy conservation and joint protection can be applied to someone with FM.

In all things, work to develop as positive an attitude as possible. That applies to parents as well as the child. Negative thoughts only reinforce pain and feeling of loss. Allow the child to grieve but not dwell on the limitations continually. Depression and anxiety seem to occur often with FM, and the jury is still out on whether it is cause or result in the development of it. At this point, it doesn't matter, what does is that you do something about that. It may be necessary for the child and the family to attend some counseling sessions. Chronic illness has a major impact on everyone in the family, not just the person who has it, and people can react in any number of ways to its presence. A counselor, particularly one who is familiar with chronic pain or chronic illness, will be able to help everyone develop coping strategies.

One area that could become a problem for the child is if he or she experiences cognitive dysfunction. This is generally evident in short term memory loss, poor concentration and difficulty in finding the proper words. Because a child's primary activity during these years requires him or her to concentrate and memorize material, it can have a major impact on school work. Find ways to reinforce what is learned in school. One of the problems with a short-term memory is that it is easily disrupted if it is not set firmly enough in the mind. See if the school will provide a tutor, or act as a tutor yourself. As a teacher I learned that the more ways an individual experiences new information, the better it sets in the mind. If the child hears material, reads the material, and writes the material, it tends to "stick" better.

Take advantage of all of the technology available through computers. If you are lucky, your school will have them accessible to all or most students. If they don't, check with the library or consider purchasing one for the family if you don't already have one. There are programs that cover a wide range of educational subjects and some schools are even offering online help.

The most important thing to remember is that while fibromyalgia will have an impact on the life of the child and his or her family, it can be dealt with effectively.

SIX

Looking Into Treatment

If you believe you have fibromyalgia, the first thing you need to do is obtain a firm diagnosis. That means finding a doctor who knows something about fibromyalgia and who is willing to work with you. It might seem as if the two requirements would be the same, but, unfortunately, there are still some doctors who know about FM and choose not to treat it or take on new cases. If that is the case, then you are better off continuing your search. Treating FM can be frustrating and time-consuming for the doctor as well as for the person with FM.

One person wrote, "I struggled so long to overcome pain. My family M.D. tried. He tried his best to treat my 'rheumatoid arthritis.' But he got tired of me as I grew more demanding for relief. I was very hurt by his diagnosis of 'chronic pain'— somehow to me, 'chronic pain' meant I was at fault. But now I know that because all the medications were useless, my doctor himself was feeling frustrated. He realized it couldn't be arthritis or bursitis, but he didn't know what it was, so to meet the insurance company's demand for diagnosis, he wrote 'chronic pain.'"

Too many times doctors do not want to admit to you, and perhaps to themselves, that they don't know what is wrong and don't know how to treat it. Even in this day, when medicine has progressed so far, there are a lot of health problems that remain a mystery. Dr. Jay Goldstein has written a book, titled *Could Your Doctor Be Wrong?* (Pharos Books), for people whose symptoms and health problems don't fit into the cookbook that physicians are taught in medical school. I highly recommend it to those who have felt they never quite fit into the nice, neat box of known illnesses.

The last five years have seen a tremendous increase in awareness of FM, with articles appearing in many of the medical journals that are published for the family practitioner or internist. I still remember how happy I was when I heard that articles about FM had finally been published in *The New England Journal of Medicine* and *Journal of the American Medical Association*. It was as if we had finally arrived, had finally been accepted as a legitimate illness. Previously, rheumatologists were the doctors most likely to recognize and diagnose FM, and their knowledge of FM is still ahead of that of doctors who practice in a more generalized field of medicine. One paper presented at the last American College of Rheumatology's annual conference concerned the performance of several different generalist and specialist resident/fellows. The rheumatologists did better at interviewing, examining, diagnosing, treating, and optimally interacting with FM patients. The researchers concluded that the more general fields—internal medicine and orthopedics—need to receive better education and training. This is not to say that you won't find a doctor in those fields who will know about FM and who will be willing to work with you. Just be aware that at this point doctors in these general fields of practice, as a whole, are not yet as familiar with FM as rheumatologists.

So how do you find the doctor who knows something about fibromyalgia and is willing to work with you? There are a number of ways to locate such a doctor. First, contact your local Arthritis Foundation chapter (see the appendix for a list of chapters in the United States). They usually provide a list of rheumatologists in their areas.

If you live near a medical university, contact their rheumatology clinic. There are several advantages in choosing medical professionals affiliated with a medical university. They may be conducting studies in which you could take part. Whenever possible, those with FM should participate in research. After all, the more research that is conducted, the better chance researchers have of finding the causes and treatment for FM. Another advantage of

1. Healing doesn't happen because of me, it happens despite me. My work is to get out of the way, to find a peaceful refuge in the midst of the pain.
2. You may be going through hell, but somewhere in your being there is ordinary peace.
3. I can't stop it from hurting. I can choose not to pay attention to it.
4. Giving up interest in pain is empowering. It permits life instead of repetition.
5. Give up any agenda for what needs to happen.
6. Don't feed the fire with fear. If it's there, identify the fear and let it be. Don't feed it, just watch it and it will dissipate.
7. As you open up, you open to a lot of strength as well as pain. If you just open to the pain, it's an unfair fight.
8. Measure progress by how calm and kind your mind remains during flare-ups, not by how much pain you feel.

—Russ

choosing a clinic affiliated with a university is the possibility of reduced fees. This is not always the case, but if you have no insurance and money is a consideration, you might check into it. On the negative side is the fact that a doctor affiliated with a school of medicine will be spending much of his or her time in teaching, working with medical students, researching, and writing. That may leave only one or two days a week for seeing patients in a clinic. Sometimes doctors who are primarily teachers and researchers may not be as good at patient-doctor relationships. I can speak from experience that this is not always the case. I believe that I was very lucky to find Dr. Rubin, and I believe he is an excellent doctor, so that I have the best of both worlds. He has access to new research, conducts some himself, and has a very good bedside manner. I took my sister to see him and she was extremely nervous about seeing this "big" doctor. Yet he put her at ease and dealt with all of her questions completely.

When a medical university has clinical facilities, medical students observe and work in the clinics under a staff physician's supervision. I have always welcomed their presence during my appointments with the doctor; this gives them the opportunity to learn more about fibromyalgia. If you are uncomfortable with the idea of medical students participating in your care, then a medical university clinic is not for you.

If there is no such facility available in your area, look for a fibromyalgia support group. Check with the larger fibromyalgia associations for up-to-date information about groups in your area. Call your local hospital or newspaper to find out if there is a group meeting near you. Very often support groups have a list of doctors who treat FM. While many groups are hesitant to openly endorse or denounce the doctors in their areas, they usually know which doctors are knowledgeable about FM and which ones will work with you. See chapter 12 for more information on support groups.

However you find a doctor, take the time to actually meet the doctor and talk to him or her about your health care and about fibromyalgia. Be prepared to pay for this visit, and don't be surprised if the doctor refuses such a visit. I called a local osteopathic physician because I wanted to get some osteopathic manipulation without having to drive to Fort Worth. I was told that a nurse would call back, but I never heard from them. This concept is still so new that many people feel uncomfortable taking this approach. You should not feel this way. You must be an active partner in your health care.

No doctor is going to have a pill that will cure your FM. Even with medicine, physical therapy, and other forms of treatment, you will most likely still have bothersome symptoms. There is only so much that medical personnel can do—the rest is up to you. After all, you live in your body 24 hours a day. You are the primary factor in your treatment. Therefore, you want to be able to work with the doctor to get best results from whatever treatment the two of you choose. FM is a chronic, life-long illness and you must be prepared to face that. By taking an active role in your health care, you are asserting some control; having control is vital to your mental well-being. Because of the nature of FM, there are some things that you may never be able to change but one of the biggest problems individuals with any chronic illness have is a feeling of being out of control and helpless. By taking an active role and working with your doctor on a planned program, you will have some control and that will be a positive step for you. Your ultimate goal is self-management to achieve the best level of functioning with pain control and improved quality of sleep possible for you with a productive lifestyle.

Managed health care is an economical reality, at least for the foreseeable future, and it has a significant impact on individuals with chronic illnesses. It is important that you find out as much as you can about your program and what it means for you, specifically. It will more than likely be more difficult to see a specialist, which is what rheumatologists are. You will need a referral from your primary care provider, if not for every visit, at least for a set number of visits. There may be other controls such as how often you may have physical therapy or whether certain medicines are covered by the insurance. Quite often, doctors are expected to keep their time with patients to a certain limit, so you need to be prepared to use that time wisely. The more you know about your managed health care program, the better you will be able to make it work for you.

Once you have found a doctor who agrees to treat you and who is experienced in treating FM, you should be able to develop a program tailored for you and your symptoms. Everyone is different and each person's case is different. Don't compare yourself with someone who has it "better" or "worse" than you. This isn't a contest. This month you might need intensive physical therapy, but two months from now you might need to increase your medication, or you might need someone who can understand the anger and depression you experience when your body refuses to let you do something you want to do. It is also likely that there will be times when you are managing just fine with less medicine or without a formal physical therapy program in the PT's office. Once you arrive at this understanding, you will have begun living "the first day of the rest of your life."

The severity of your symptoms and the needs of your lifestyle will determine the means of treatment. It may be necessary to try a number of drugs, modes of treatment, or combinations of modalities to find what works best. Don't expect a miracle cure. There isn't one yet and spending your time wishing for one prevents you from achieving progress in areas you can do something about. Mild medication enables some people to sleep better and they see an improvement almost immediately. For others, it may take a period of hospitalization to break the pain cycle. Although many of those seeking a diagnosis for fibromyalgia have spent a lot of time in hospitals during their search, once they get a diagnosis, their frequency of hospitalization decreas-

What Your Doctor May Do and Say

1. Reassure you that the syndrome is a real medical entity but is not progressive, crippling, or life threatening.
2. Point out that prospects for medical cure are only moderate.
3. Identify and treat concomitant painful conditions, psychological disturbances, or other factors that may exacerbate or contribute to symptoms.
4. Suggest positive environmental changes.
5. Encourage you to:
   • take an active role in the management of the disease;
   • correct stresses to the cervical and lumbar spine by the use of neck-support pillows during sleep and abdominal exercises to strengthen muscular support of the back;
   • begin a gradually progressive program of aerobic exercise to improve the level of muscular fitness and sense of well-being;
   • remain physically and socially active;
   • identify and eliminate stresses or environmental disturbances that may exacerbate sleep disturbance and symptoms;
   • avoid using the diagnosis for secondary gain, such as avoidance of unwelcome tasks;
   • avoid overdependence on medical personnel or resources.
6. Prescribe low doses of tricyclic medications to improve sleep quality.
7. Prescribe simple analgesics, such as acetaminophen, as needed.

Courtesy of Dr. Robert M. Bennett. Reprinted with permission from *Patient Care.*

es dramatically. Treatment may range from regular appointments to the doctor to an extended pain management program.

If you already have a doctor who is treating your fibromyalgia, ask yourself if you are happy with your relationship. Of the people I have surveyed, about 25 percent were not satisfied with their doctors. Many of them felt they knew more about FM than their doctor did or indicated their doctors never took the time to talk with them. However, it is encouraging that almost 50 percent of the group felt their doctors were knowledgeable and caring. Their doctors took the time to sit down and talk with them about fibromyalgia, its impact, and its treatment. This is a great improvement from the time when a "good patient" was one who asked no questions and left everything to the doctor. This style of doctoring has almost disappeared, as both doctors and patients realize that the physician is human and also has good and bad days. You can help facilitate excellent care by carefully writing down any questions you may have or anything that you do not understand. This helps both of you address what you need to know during the office visit.

In the research literature on FM, authors consistently encourage doctors to take the time to discuss FM with their patients and to reassure them. This assurance and your active involvement provide the foundation for living with FM. In fact, almost all of the prominent researchers and leaders in the field list education as the first step of a treatment plan. The more you know, the better you understand what is going on with your body, the better able you are to cope positively with it.

Whatever other specialists your doctor involves, you will often find that it takes all of them to produce some advances in your treatment. Treatment must be multidisciplinary. We are made up of body, mind, and spirit, and modern medicine is finally recognizing that they are so intricately intertwined that it is impossible to separate them. We will look at the treatments that affect the body first, then the mind, and lastly, the spirit.

Okay, now you have a doctor and you know something about FM so it's time to set up your game plan. What kind of drugs are available? What about physical therapy? How can you control your pain? Just how effective is a particular mode of treatment? Remember, you are striving to break the pain cycle, to obtain whatever relief the medical profession can give you, and then to make changes in your lifestyle that will extend the relief.

At this time, there is no one treatment that will improve the symptoms. Because the cause of FM is still unknown, it is only possible to treat the symptoms. Studies have shown that a multidisciplinary approach is necessary and is the most effective. Although your daily care does not need quite the exact control that a scientific study does, you need to have an idea where you are, where you are going, how you are going to get there, and how to tell if you're there. Confused? Don't be. Just think of it as a journey; after all, life is a journey, not just a goal. Your progress with your own fibromyalgia and its symptoms are your journey and the objectives are the signposts along the way. The objectives are usually to improve sleep, reduce pain and stiffness, and increase your functioning.

Until lately, no one thought a lot about how to measure your progress although the goal might have been to get, say, a 50 percent decrease in your pain. But, how do you really measure pain? There's no lab test that says, okay, you've decreased it by 25 percent and you still need to reduce it by another 25 percent. FM isn't like high blood pressure, where you can take a measurement and see improvement, or diabetes, where your blood levels indicate whether you are improving or not.

Several researchers have been working on ways to measure pain, as we discussed in an earlier chapter. At this point, no one instrument has been accepted as a standard measurement for pain but there are several fairly accurate tests that are being used. One involves shading a figure to indicate just where an individual experienced pain in a recent time frame; these areas were totaled as a percentage of the whole body, much as burns over the body are figured. Another method uses a vertical line scale with the individual marking somewhere along that line the level of pain; the left end indicates no pain and the right end an extreme pain or "the worst pain ever." Some doctors add descriptive phrases along this line to help an individual be more accurate. Other doctors use questionnaires that don't take more than a few minutes to complete and which give an indication of how the individual is sleeping and how active he or she is.

Pain management is a relatively new field in medicine. In the past, physicians focused on the use of narcotics in the treatment of acute pain, but they were warned in medical school to beware of allowing a patient to become addicted to drugs. Most often, the dosage given is much less than what is necessary to alleviate the pain.

There are two basic reasons for the difficulty in treating chronic pain; the lack of understanding the way pain is felt by the patient and the difficulty in measuring pain. In recent years, more researchers are studying pain pathways and the best way to treat acute pain. Treatment of chronic pain has shown less progress, but it has become obvious that there is no single method for relief. Fibromyalgia has become almost the standard in chronic pain conditions.

Part of the difficulty has been the differences in how individuals experience pain. There are almost unlimited variables that combine to determine an individual's perception of pain. These factors can range from genetics and disposition, to an ethnic culture that teaches a person whether it is okay or forbidden to verbalize pain, to various aspects of the central nervous system.

One thing that has become obvious from studies on the various treatments of fibromyalgia is that there is no one single regimen that will bring about significant relief. To date, nothing, not drugs by themselves or in combination with physical therapy, exercise, and cognitive behavior adjustments, has brought about total long-term relief. But combinations of these and other methods have brought some temporary improvements in symptoms.

So, although we break the treatment down into various modalities, remember that it usually takes a combination of methods to be effective. These combinations may have to be changed as needed. The four areas to be considered in treatment are pain, exercise, sleep, and the psyche. The standard three methods are medicine, which is used to improve sleep because sleep has been shown to be directly related to decreasing the pain level; physical therapy, which includes working with the muscles by stretching and aerobic exercising; and a cognitive behavior modification program, which includes stress management and coping strategies. A person may take a combination of a tricyclic antidepressant; a nonsteroidal, anti-inflammatory drug; have a short-term physical therapy course, including mild aerobic exercise; and learn relaxation therapy. Such combinations will be individualized and should be altered as needed.

Go into treatment with a realistic attitude. You and your doctor may set goals such as a 50 percent improvement in pain level or 25 percent improvement in sleep. Of course, these figures are arbitrary; only you and your doctor can set realistic goals.

Long-term studies have now followed some individuals for as long as 15 years. Essentially, their condition remains the same. There have been fluctuations based on weather conditions, time of the year, physical activity, amount of stress, and even the time of day, but in the long run, individuals did not get significantly worse and some even improved. A few experienced a complete remission of symptoms. Usually, the best rate of improvement was among those whose FM had been diagnosed early and treatment begun within a short period of time after symptoms appeared.

You should keep a record of all of your medications, with the amounts and schedule of when to take them. Nurses love it when they ask what I'm taking and I can provide them with a written card that I carry around with me in my purse. I try to keep it up to date, reflecting any changes that we make. It also helps in times of emergencies, when paramedics or emergency room personnel need to know what you're taking. Many individuals with FM are sensitive to certain drugs, and this record can be an important aid in determining a safe course of emergency treatment.

It is also a good idea for you to keep a journal of your treatment regimen and symptoms so that you can look back over it and see that "yes, you have improved," or "no, you haven't". I do keep a record of my medicine but I prefer to not dwell or focus on my pain. I listen to my body and do my best to act accordingly. Many doctors recommend a journal or a record of some sort so that you can check your progress.

Medication

Until the cause or causes of fibromyalgia are known, treatment must concentrate on alleviating the symptoms. Those symptoms include the widespread pain, muscle spasms, and sleep disturbance. Sometimes by treating the sleep disturbance, the pain and muscle spasms can be relieved. Different drugs from several categories are prescribed for those with FM. Because of the hypersensitivities that someone with FM may have, any medicine should be taken cautiously. Start with small dosages and slowly increase to the level your doctor recommends, while watching for side effects. To accurately judge their effectiveness, some medicines must be taken for about 1 month or so, but if the side effects are too severe or appear potentially dangerous, check with your doctor.

Analgesics
The use of such simple analgesics as aspirin or acetaminophen can relieve most of the pain for FM patients and enable them to sleep more comfortably. For 6 years I was able to control my pain with these over-the-counter drugs. Several people I surveyed indicated that these are all they need except in

special situations. Some take coated aspirin to prevent stomach upset. Be aware that there is generally no difference between a brand name and a generic product. Your doctor can recommend the amount you should take to control pain; if he or she doesn't, begin with a reasonable dose and take it on a regular basis. If your pain and fatigue is enough to interfere with your normal functioning, Dr. Glenn A. McCain recommends that you take acetaminophen regularly around the clock instead of on an "as needed" basis. He feels that the pain-contingent (or as needed) schedule pairs two stimuli (the pain itself and the analgesic), which allow the nervous system to become conditioned to ongoing pain.

In chronic pain, the messages that are being sent to the brain are in error, and as some researchers believe the nerves and the brain, have become overstimulated, over sensitive. You don't want to add to that stimulus. Many doctors recommend that you stop using caffeine, as well as alcohol, and tobacco. Caffeine is found in soda, tea, coffee and chocolate. Be sure to read the contents on anything you put in your mouth; caffeine is added to some things you wouldn't expect.

Nonsteroidal Anti-Inflammatory Drugs

Because FM is a chronic syndrome, pain relief from drugs must be handled carefully. An addiction to painkillers will only complicate matters. If relief can't be obtained from either aspirin or acetaminophen, it is usually sought from nonsteroidal anti-inflammatory drugs (NSAIDs), many of which are now available over-the-counter. There is no inflammation in FM as there is in other rheumatic conditions, so these drugs are looked to more for their painkilling effects than for their anti-inflammatory action. Just as in other rheumatic conditions, they are often prescribed on a continuous basis in an attempt to establish a certain level of the drug in the bloodstream.

Generally, the range of nonsteroidal anti-inflammatory drugs is tried in an attempt to find one that works best for an individual. The major problem with these drugs is not potential addiction but the damage they can do to the stomach. Unfortunately, few drugs have no side effects, so we end up treating the major health problem (pain) and then treating the side effect of the painkiller (irritated stomach or an ulcer).

1. Depending on the medicine, ask your doctor if he can prescribe a higher dosage and then cut the pills in two; this saves me $25 a month on one medicine.
2. Buy a light-weight canvas (about $7) camping "chair." Put it in your shopping cart and whip it out when you get tired or in pain and rest anytime you want.

—Dorothy

Some NSAIDs are naproxen (Naprosyn), piroxicam (Feldene), ibuprofen (Motrin, Advil, Nuprin), indomethacin (Indocin), sulindac (Clinoril), diflunsal (Dolobid), nabumetone (Relafen), and tramodol (Ultram).

Narcotics

The use of narcotics (propoxyphene, codeine, ocycondone) is still very controversial in treating all chronic pain, not just fibromyalgia. Although many doctors are afraid of addiction, there are others working with chronic pain, who say what might occur is not addiction but a physical dependence. There is a difference. Many cancer patients have been left in severe distress because the doctor was concerned about addiction. For terminal patients, the question should be moot.

Physical addiction can occur when an individual takes a particular drug for more than just a few days. If the person suddenly stops taking a drug, such as morphine, rather than slowly decreasing the amount taken until they stop completely, they will probably suffer temporary "withdrawal symptoms" such as nausea, muscle cramps and chills. But addiction occurs when an individual has a constant craving for the drug's euphoric or calming effects and will do almost anything to obtain the drug, even if it is self-destructive. I don't know about everyone else with FM but for myself and several of the individuals I have talked to who have it, there is no euphoric or calming effect. In other words there is no pleasure in taking it, only the relief of the pain. There are only a very few of the stronger pain-killers that I can tolerate without nausea or hallucinations or other unpleasant side effects. Several years ago, after I experienced both nausea and nightmarish hallucinations from taking one particular drug, I told my doctor I would rather have my pain. I count myself lucky that I can take Tylenol with codeine #3 or Toradol.

For those with chronic, non-terminal illnesses that leave them in severe pain, the question isn't so simple. There has been a lot of focus on chronic pain treatment in the media lately. Too often doctors who have dared to work aggressively in maintaining a bearable level pain for their patients have been pulled before a state review board and had their license to practice medicine revoked. Even pharmacists filling the prescriptions have had problems. But there are efforts being made to better understand the problems with chronic pain and recognize that not everyone who takes narcotics for pain relief will become addicted to them. Generally, those who take them as needed do not crave more than they need.

With that said, it must be emphasized that any narcotic painkiller should be used conservatively and only when necessary. There are times, as in a flare-up, when it takes the strength of a narcotic to relieve pain. It is better to stop pain when it begins instead of letting it build until it is truly unmanageable. If your doctor is willing to prescribe one of these drugs for you, make sure you use it wisely. If you can get pain relief from some of the other medicines and methods we've discussed, do so. I tend to think of using my Tylenol with codeine #3 (acetaminophen with codeine) for minor pain as comparable to using a sledgehammer to kill a fly—overkill.

Tricyclic Medications and Other Antidepressants

Because sleep disturbance has been indicated as a source of increased pain and fatigue, medications that improve sleep are used. Initially, Dr. Harvey Moldofsky's sleep studies seemed to indicate that the sleep disturbances were tied into the fatigue and chronic pain. This idea led to seeking drugs that would increase the stage four, or non-rapid-eye-movement (NREM) restorative sleep. *Tricyclic antidepressants* such as amitriptyline (Elavil or Endep), have been shown to have an impact on the central and peripheral nervous system and early studies demonstrated a positive response to amitriptyline in people with FM on a short term basis.

However, follow-ups on many of those who took part in the study found that positive response was lost over a period of time, and that the symptoms eventually returned to their original level. There are still confusing results from studies on many of these drugs, probably because we are dealing with a number of subgroups of those who have FM. Everyone is not the same and therefore, studies often come back with different results. Until we learn to control every variable (and that will be next to impossible, given that every individual brings a different set of personal traits, beliefs, and experiences to how they deal with pain) it will be difficult to duplicate exact reactions to both the pain and the medication. When researchers are able to recognize subgroups of patients who have a great deal in common with each other, research study outcomes will improve.

Other drugs in the same family have not proved to be as effective in the trials but may work for an individual. These are nortriptyline (Pamelor), imipramine (Tofranil), clomipramine (Anafranil), and maprotiline (Ludiomil).

(Some believed that the fenfluramine-phentermine (fen-phen) combination was a successful treatment for FM. These drugs were primarily used for weight loss and neither drug was approved for use in FM. Fenfluramine was pulled from the market in 1997 because of dangerous side effects and is no longer available for prescription, although phentermine (Adepex) is still available. Be cautious about unproven drugs that are advertised as a "cure" for FM. The Internet (World Wide Web) is a source of much health care information, but make sure that that source is a valid and reliable one. We will discuss alternative treatments and the Internet later in the book.)

Tricyclics can reduce alpha-wave intrusions into stage four NREM sleep, block the re-uptake of norepinephrine and serotonin (chemicals that help transmit pain signals), and increase endorphin stimulation levels. The dosage used in the treatment of fibromyalgia is far lower than that normally given for depression. Quite often, when a physician suggests the use of a tricyclic antidepressant, the person with FM is quick to conclude that the doctor believes the problem is psychological. Dosage for chronic pain is generally no more than 10 to 40 mg per day, depending on the drug, while that for depression can be as high as 150 to 200 mg per day.

Because of the nature of the symptoms of FM, it is hard to prove a drug's contribution to improvement. A study conducted at the University of Alberta by Dr. Roger Scudds and Dr. Glenn McCain, who is from the University of Western Ontario, sought improvements in "outcome measures of pain, tender point sensitivity, and patient assessment of well-being." Their results showed improvement in overall myalgia as well as pain threshold and tolerance. The overall myalgia improvement came after the active treatment period, when the tender points became less responsive to pressure.

Tricyclic antidepressants remain the top medicines of choice for lack of better drugs, although there are an increasing number of studies on the newer class of antidepressants such as fluoxetine (Prozac), paroxetine (Paxil), and sertraline (Zoloft), which are selective boosters of serotonin availability in brain nerve junctions. They are sometimes given together with one of the tricyclics.

One last note here, although we will discuss it more later. Many studies have found that individuals with fibromyalgia are often also depressed and anxious. No one knows yet if it is as a result of the FM or simply another aspect of the associated conditions that we discussed earlier. The important thing is that depression should be treated as well as the pain, sleep disturbances, and other problems. Studies have found that if these two concerns are treated, the pain and sleep disturbances improve. So don't think your doctor believes that your pain is psychological—he just recognizes that depression is one more factor that has to be addressed.

All medicine has some side effects, even aspirin, but many individuals find they have problems tolerating tricyclic antidepressants. Common side effects include dry mouth, drowsiness, increased weight, fluid retention, difficulty in urination, and sometimes a "medicine hangover" the next day. The first thing to remember is that your body may be trying to catch up on lost sleep and it isn't unusual to sleep around the clock when first taking Elavil or one of the others. Hard candy or chewing gum helps with the dry mouth and the hangover might be avoided by taking a lower dose or taking it an hour or two earlier in the evening.

Muscle Relaxants

Cyclobenzaprine (Flexeril) is another member of the tricyclic medicines, but it is primarily a muscle relaxer. It may affect the central nervous system as well as providing peripheral actions. Dr. Robert Bennett et al. conducted a trial study comparing cyclobenzaprine and a placebo on 120 patients with fibromyalgia. The results showed improvement in the quality of sleep, a decrease in pain severity and level of fatigue, but showed no improvement in morning stiffness.

Other muscle relaxers sometimes prescribed are methocarbamol (Robaxin), chlorozo-xazone (Paraflex), chloroxazone (Parafon Forte), orphenadrine citrate (Norflex) and carisoprodol (Soma).

Benzodiazepines

Benzodiazepines are anti-anxiety drugs that have been helpful in reducing muscle spasms, pain, and sleep disturbance. Alprazolam (Xanax) is perhaps the most commonly used. It has been studied in combination with ibuprofin, where it produced improvements for some people with FM. Numerous studies have shown that many people with FM face anxiety and depression, and that compared with those with other chronic illnesses such as rheumatoid arthritis or lupus, they experience more daily hassles. If the primary problem with FM is indeed an improper response to stress by the body's sympathetic and parasympathetic systems, these daily hassles would constitute stimuli to which the system overreacts. Whatever the cause, anxiety and stress have a major impact on quality of life. Taken carefully and under your physician's direction, anti-anxiety medications can prove a useful addition to your overall healthcare program.

Summary

Although all these drugs provide some relief for some people with fibromyalgia, none totally relieves all symptoms. Yet you shouldn't be pessimistic when starting out. Be willing to give your doctor and these drugs a try. You may end up like the 54-year-old teacher who was diagnosed in 1990 with FM and who has obtained relief with piroxicam (Feldene), a nonsteroidal-anti-inflammatory drug. She says that her activities are not limited at the present time and that Feldene has almost completely relieved all her pain. Another point to make—trials with new and different medications are underway, and we may reap the benefit of those researches with relief. Don't give up hope, even if the standard drugs covered haven't provided the level of relief you want.

Learn as much as you can about any medicine you take. Know what drugs your doctor has prescribed for you, how to take them, and what their side effects may be. Never let anyone (a doctor or nurse in an emergency room, for example), give you any medicine with which you are unfamiliar. In the past, healthcare providers didn't believe in telling you what they were giving. The nurse would only say, "Doctor ordered it." After experiencing some severe reactions, including respiratory difficulties, I adopted the rule, "No knowledge, no medicine." After all, this is my body we're talking about, and I'm the one who will experience any difficulties that may develop. Luckily, things have changed enough that I get the information without any problems.

Make a point of telling each of your doctors, if you have more than one, what other medications you are taking. Drug combinations can create serious side effects. If you don't tell your cardiologist what your rheumatologist has prescribed for you, and vice-versa, you may experience serious problems. Buy one of the excellent guides to prescription and over-the-counter drugs and keep it handy.

If you can't read your doctor's prescription, make sure the pharmacy writes the generic or brand name of the drug on your bottle as well as clear instructions for taking the medicine. This is usually required by law. Most pharmacies must counsel individuals regarding the medicines they dispense. Providing a computer-generated printout is part of that counseling. (Another source of pharmaceutical information is the Internet, which we will discuss in more depth later.) Always ask your doctor if you have any questions and don't suddenly stop taking any medication without first discussing it with your physician.

SEVEN

Options for Physical Treatment

The second leg of the treatment program involves any of a number of non-drug modalities. Physical activities and modes are an important aspect in this area of treatment. Many people find they can get significant relief from physical therapy and other hands-on assistance. In a number of studies, it has been proven that a combination of modalities works best. Explore the non-drug treatments that are available to you.

Physical Therapy

Physical therapy for people with FM includes both passive and active modes. According to Katherine B. McCoy, of the West Portland Physical Therapy Clinic, and speaker at one of the conferences on fibromyalgia, there are three procedures used to treat the FM patient.

• Assess musculoskeletal status including posture, strength, range of motion, ligamentous stability, endurance, and muscle tone

• Directly manage FM patients' exercise programs with the goal of reaching optimal fitness and increased endurance

• Increased strength should be attained while maintaining decreased levels of pain and fatigue via pain modulation techniques.

The main objective of physical therapy is to ease muscle spasms through relaxation, thereby easing pain. The second objective is to help you learn normal neuromuscular functioning. Physical therapy is done on either an inpatient or an outpatient basis. Hospitalization is usually necessary only when the pain is so severe that a combination of physical therapy and bed rest is required to break the pain-spasm-pain cycle. But most PT can be done on an outpatient basis. Any physical therapy program should be considered for a limited time frame. Just as you cannot live in your doctor's office, neither can you live at the physical therapist's. The idea is to help you reach a maximum level of functioning with a program you can carry on at home.

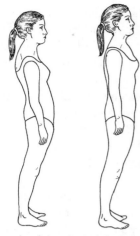

*One role of the physical therapist is to check posture as it has an impact on the muscles. If one thinks of the body as a structure requiring complete balance, slumped shoulders or a swayback stance as shown **on the left** puts that structure out of balance. This then, puts a strain on the muscles which can't do their proper job and, therefore, become stressed and painful.*

Although new advances are being made daily in the study of pain, the chronic pain of fibromyalgia has proved difficult to understand. It seems pretty clear at this point that while the major problem is in the central nervous system with pain processing and perception, the muscles are involved as well. In numerous studies, it has been found that the muscles are not just in spasm as was once thought, but that there are problems with blood flow, muscle tension (some people with FM are unable to relax certain muscles after they have been contracted), and increased levels of calcium after exertion. Most physical therapy is directed toward the muscles, to relieve the spasms and tightness, but many times these activities will also be effective at the central nervous system level as well.

Relaxing the muscles begins with the application of either heat or cold. This usually involves the use of hot or cold packs applied to the painful areas. Many people feel better after the hot packs are applied, while others cannot tolerate the heat. Everything must be tailored to the individual; there is no broad overall treatment strategy that works for everyone. There are heating pads, providing either dry or moist heat; hot or cold packs, placed in the microwave or the freezer; mechanical massagers; infra-red massagers; and bathtub whirlpools and hot tubs that can be used at home. Use common sense in using any of these and always check with your doctor or physical therapist first.

Ultrasound or gentle massage following hot packs is another form of treatment. The hot packs provide superficial heat, while ultrasound sends high-frequency waves into the muscles, ligaments, and fascia from a small hand-held device, causing heat to penetrate deeply into the muscles. Some studies suggest that, in the future, laser devices may be used to eliminate the waste products within the muscle cells.

1. *Laugh daily, for laughter is the best medicine!*
2. *Volunteer. Yes, there are people less fortunate than you.*
3. *Self-care. Gentle and loving thoughts and touch.*

—Pamela

Electrical therapies can also be done in the physical therapy department. These include high voltage galvanic stimulation, micro-am stimulation, and transcutaneous electrical nerve stimulation (TENS). Some individuals find that portable TENS units provide them with temporary relief. The concept of TENS is that it sends an electrical impulse along the nerve pathways, moving faster than the pain message. Because the brain can only register one signal at a time, it experiences the electrical impulse and ignores the pain message. Unfortunately, after a period of time, the brain adapts and overcomes the treatment on the same spot. Unlike portable units, those used by PTs cover large areas, and because they are not used constantly, the brain doesn't overcome them as easily.

The second phase of physical therapy involves an exercise program that targets increasing range of motion, strength, and endurance. Several studies have shown that an aerobic exercise program is best for those with FM. Aerobic exercises are those that improve cardiovascular fitness, or the flow of oxygen in the blood throughout the body. People with fibromyalgia are often hesitant to exercise, since experience has taught them that pain will follow exertion.

Whether in a physical therapy setting or at home, a mild stretching program should follow the application of heat and should always be done before any exercise is attempted. This helps get the muscles loosened up so they will move more easily and there will be less danger of injury to the muscle. When the stretching has been tolerated well, a very gradual active exercise program is begun. It is not uncommon to begin with simple range of motion exercises such as those recommended by the Arthritis Foundation.

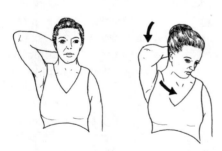

Although not used by everyone, some physicians recommend a spray-and-stretch technique. A vapor coolant, such as ethyl chloride, is sprayed over the tender area to anesthetize it, and then the physical therapist gently stretches the muscles.

Strengthening exercises include isometric, isokinetic, or isotonic exercises. Katherine B. McCoy recommends starting with just lifting the weight of your own arm and then work up to lifting a six-ounce can of soup. Most of the larger cans weigh one pound. Your physical therapist may recommend other ideas for strength-

Before beginning your aerobic exercises, you should work through a series of range of motion exercises, slowly and gently to warm up your muscles. One of these is the lateral neck stretch. Place one hand at the base of your skull and slowly turn your head until you are facing along the inside of the opposite arm. Take each set of muscles through their range of motion, if pain follows, move more slowly and don't force the movement. Instead, try each day to improve on the range of movement, but don't stop the exercises and stretches completely.

General Principles of Exercise

1. Exercise on a daily basis.
2. To minimize stress on your joints, exercise periods should begin with a period of gentle stretching or warm-up activities and end with a cool down period.
3. Prepare your body for exercise. Wear appropriate clothing and shoes for exercise.
4. Exercise when you have the least pain or stiffness, when you are not tired, and when your medication is having the most effect. If joint pain or stiffness is a problem, you may massage the painful area or apply heat or cold to the area.
5. Stay within your limits. Don't try to "test" yourself by choosing an activity that is too hard or too stressful for your body.
6. Perform exercise in a slow, steady rhythm allowing muscles time to relax between repetitions.
7. Breathe in a normal deep rhythmic pattern, coordinating breathing with exercise. Do not hold your breath while exercising.
8. If your joint is hot or inflamed, care needs to be taken with exercise. The inflamed joint should be rested and only moved gently through its range of motion. (If you also have arthritis.)
9. Avoid exercising on a full stomach if you find it uncomfortable.
10. Begin your exercise program with a few repetitions of each exercise. Gradually increase the number of repetitions as your range and strength improve.

Offered by Arthritis Care Center, Saint Joseph Hospital, Fort Worth, Texas.

ening the muscles but care should be taken to avoid any eccentric exercises that tend to aggravate muscle pain. In moving, the muscles of the body contract and lengthen. When the elbow is bent, the biceps muscles contract, which is concentric contraction. If the muscles are then forced to lengthen, as occurs when pushing a vacuum cleaner or when reaching overhead to put dishes away, eccentric contraction occurs. It is this eccentric contraction that causes pain in those with FM.

Sharon Clark, of University Health Science Center, Portland, Oregon, is one of the leading experts on exercise and fibromyalgia. She describes how to decrease eccentric work by using the concept of the old-fashioned hoop skirt. By picturing the narrow waist of the hoop skirt at the neck level, draw an imaginary triangle with two sides out and down and the third cutting across the knees. By keeping the arms within the triangle, the individual prevents eccentric work of the upper body. Clark also recommends taking smaller steps when walking downhill and downstairs to keep eccentric work of the lower body to a minimum.

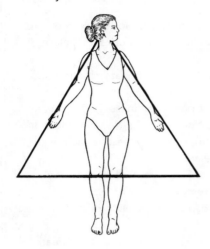

Dr. Sharon Clark borrowed the concept of the old-fashioned hoop skirts to give those with FM a guideline about range of movement in exercises. She has found that the trouble comes when a muscle contraction is released that FM affects that muscle. So she recommends that you try to keep your movements within the "hoop".

In a handout given to participants at an FM conference Katherine McCoy also recommends "a spinal stabilization program for FM is necessary since spinal asymmetry and poor posture as a result of deconditioning, fatigue, and depression are prevalent." The human body was designed to be a perfectly balanced structure, with the musculoskeletal system working to maintain that balance. If muscle pain occurs, the individual will often develop an "unnatural" position in order to protect the muscle or ease the pain. If a person wears a pair of ill-fitting shoes that causes them to shift their body weight somewhat, the muscles must work to counteract that shift. These or any of a number of other factors can throw the natural or proper posture out of alignment. The physical therapist must then work to bring the spine back into its proper position by any of several therapies, from range of motion movement and stretching to exercises which will strengthen the muscles.

Once you have warmed up, you can then move on to such aerobic exercises as walking, swimming, or bike riding. Aquatic exercise is also an option as long as it is all done underwater and in a heated pool. The water reduces the stress of weight on the body. Just be careful not to get into a program that is too aggressive and where many of the upper body exercises are done out of the water. That is counterproductive to what you need.

Aerobic exercise improves your cardiovascular fitness and aids in the body's production of its natural endorphins and hormones. Not all researchers agree upon how such a program should be conducted. Some agree with Dr. Paul Reilly and Dr. Geoffrey Littlejohn who believe that although this may make the symptoms worse at first, you should push into the 'pain zone' because you will do yourself no harm by doing so. However, they do suggest that you start gently and persistently increase the intensity of the exercise. Others believe that pushing too hard and attempting to go through the pain will only increase the likelihood that the individual will drop out of the program. Rather than losing time because of pain, they recommend that you move very slowly and increase your time and repetitions at a very gradual rate.

The key is to start gently and increase gradually. If you have been accustomed to an active lifestyle, it is sometimes hard to realize that the most you can start your exercise program with is five repetitions of arm movement or two minutes on an indoor bicycle. That is so little that health-conscious people may question the benefits of doing anything if you must start so low. But once you have managed to ride the indoor bicycle for two minutes at a time, perhaps twice a day for several days, you will find that you can gradually increase your time.

During my initial hospital stay in 1987, I worked up to three repetitions of some stretching exercises and four minutes on the bicycle. One day I decided that I would increase my repetitions and increase the time on the bike to ten minutes. That night I was in a lot of pain. As a result, I had to return almost to the beginning stages of my therapy. Be reasonable in your expectations for progress.

Usually aerobic exercise and its progress are measured by your maximum heart rate. To determine your maximum heart rate (MHR) subtract your age from 220. I am 50 years old, so my rate would be 170 beats per minute. To estimate the intensity of your training prgram, follow Sharon Clark's chart for common activities.

Normal Activity	0.50
Leisurely walking, cycling, dancing	

Low Intensity	0.60
Sustained activity with no more than slight sense of exertion; walking, cycling, dancing	

Moderate Intensity	0.70
Can talk when exercising, feels some exertion; brisk walking, jogging, aquatics	

Exercise above this level is not recommended.

Several forms of aerobic exercises, including walking and bicycling, can be started at a low level and gradually increased. Exercise is important for a number of reasons. If followed regularly, it brings the body back into a state of conditioning and stimulates the body's production of endorphins. Moderate exercise has also been proved to improve stage four sleep. If you find that you can't walk around the block or if the weather makes it too miserable for you, you can still work something out. I live in a 76-foot mobile

Simple Range-of-Motion Exercises

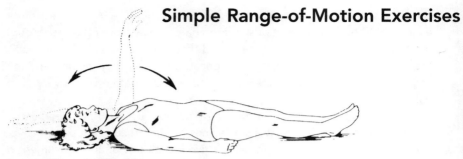

Figure 1. Shoulder
Lie on your back. Raise one arm over your head, keeping your elbow straight. Keep your arms close to your ear. Return your arm slowly to your side. Repeat with your other arm.

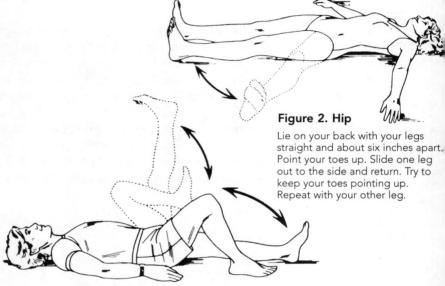

Figure 2. Hip
Lie on your back with your legs straight and about six inches apart. Point your toes up. Slide one leg out to the side and return. Try to keep your toes pointing up. Repeat with your other leg.

Figure 3. Knee and Hip
Lie on your back with one knee bent and the other as straight as possible. Bend the knee of the straight leg and bring it toward the chest. Push the leg into the air and then lower it to the floor. Repeat, using the other leg.

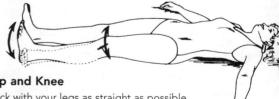

Figure 4. Hip and Knee
Lie on your back with your legs as straight as possible, about six inches apart. Keep your toes pointed up. Roll your hips and knees in and out, keeping your knees straight.

To further strengthen knees, while lying with both legs out straight, attempt to push one knee down against the floor. Tighten the muscle on the front of the thigh. Hold this tightening for a slow count of five. Relax. Repeat with the other knee.

Figure 5. Shoulder

a) Place your hands behind your head.
b) Move your elbows back as far as you can. As you move your elbows back, move your head back. Return to starting position and repeat.

Figure 6. Thumb

Open your hand with your fingers straight. Reach your thumb across your palm until it touches the base of the little finger. Stretch your thumb out and repeat.

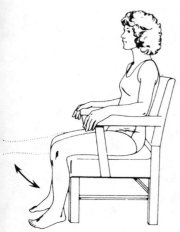

Figure 7. Knee

Sit in a chair high enough so that you can swing your leg. Keep your thigh on the chair and straighten out your knee. Hold a few seconds. Then bend your knee back as far as possible. Repeat with the other knee.

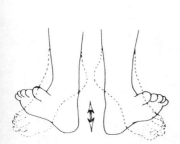

Figure 8. Ankle

While sitting: a) lift your toes as high as possible. Then, return your toes to the floor and b) lift the heels up as high as possible. Repeat.

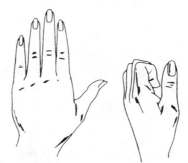

Figure 9. Fingers

Open your hand, with fingers straight. Bend all the finger joints except the knuckles. Touch the top of the palm. Open and repeat.

After coping with fibromyalgia for 13 years, I have found the following work best for me:
1. *Exercise; walking, or biking.*
2. *Yoga; mind and body relaxation techniques.*
3. *Low fat diet and daily vitamin supplements including multivitamins, V.T.E., and calcium/magnesium.*
4. *Minimize stress.*
5. *Get plenty of sleep and take small doses of Flexiril when needed.*

—Josey

home and because I haven't been able to get very far walking outside, I walk inside, from one end of the mobile home to the other. If I'm having a bad day, I don't have to walk as far or make as many trips from one end to the other.

I cannot walk very far without my pain level rising drastically and my energy level running into the red. I always felt guilty that I couldn't do better and sometimes thought that Dr. Rubin was disappointed in me, until the day I arrived in his office after having to walk up the hill from the parking lot two blocks away because there were no handicapped spots open. I was in pretty rough shape, struggling to get my breath and with my blood pressure elevated because of the pain. He saw me as he was moving between examination rooms and insisted that I was to park close to the door and let them know I was there. They would either get a wheelchair and help me get to the office or they would park my car for me.

His office has since been moved to another location at the medical school complex with plenty of handicapped parking spaces and I now have an electric scooter. I use the scooter when I have to go any distance (my concept of distance) but I also make it a point to walk if I can manage, such as going into a convenience store or my local grocery store. Although these short trips are not part of a regular exercise program, it is exercise and activity. Even though I use that scooter, I know that if I don't walk as much as I can, my muscles will atrophy and then I might not be able to get even that far. So, even if it seems minimal, develop some sort of exercise and activity.

Once released from the hospital or from the physical therapy clinic's outpatient program, it is up to you to continue to practice what you have been

Finding a physician knowledgeable about fibromyalgia changed my life around. I am now a happy functioning human being again. I obtained his name from FM Network in Arizona. He advocates a good night's sleep, massage adjustments, exercise, and medications specific for me. —Delene

shown. It isn't easy to apply hot packs to yourself, but a warm tub bath or regular sessions in a whirlpool or hot tub can easily take their place. You might try using a moist heating pad on low for about 30 minutes. (Intense heat may drain your energy. If you use a hot tub, it is recommended that you don't stay in the water for more than 5 to 10 minutes if the temperature exceeds 104 degrees.)

Once you are no longer supervised, it is all too easy to let the exercise slide, even if it's just a day or two. You may feel like there's so much else to do. You're not alone in this behavior. There are indoor bicycles, exercisers, and treadmills gathering dust across the country. But you must push yourself to stick to your exercise schedule because it will take you longer than it does other people to regain past improvements. And, of course, there is also the chance that you will give up exercise entirely.

> *Moderate- to low-impact exercise at least 5 times a week lessens aches. Don't skip exercise two consecutive days. When I do, the aches get intense. I've found Ultram to be an effective pain killer. I've become accustomed to it and can be productive while taking it. —Holly*

Jacqueline King was a physical therapist for 35 years, including 8 years in her own private practice before retiring. She has seen many people with FM and has watched them struggle to find beneficial treatment. "No one wants to take it on," she says. "Too many doctors don't want to spend the time that is necessary, but even beyond that, the healthcare field is affected by the economics of long-term care. I know this is an extreme case, but I would consider it on the same level as treating cancer. No one is going to die from FM, but they must realize that it is forever. There is no physical therapy or other form of treatment that will cure it or bring total relief.

"There must be a total change in thinking, in lifestyle, and most people are afraid and resistant to change. There must be a commitment to that long-term treatment, not only by the individual but also by the doctor and the healthcare field, including insurance companies. But there is no guarantee that treatment will bring about relief, and that lack of guarantee leads many doctors to not want to get involved as well as individuals to say 'It's no use. It won't do any good, so why even try to exercise?'."

Having helped a number of people with FM, Ms. King believes individuals must approach their situation from the right perspective. She contends that those with FM face enough negativity from the medical community without contributing to it themselves—even though it is easy to become self-defeatist under the circumstances. Ms. King says, "So much is left to the individual. There's not a lot we can do about the health field, although I believe some changes are going to be made to provide care for those who don't have insurance. But people must first decide they are going to do whatever they can to change their lifestyle and to include exercise. I have to admit that most won't do so on their own. That's why I think they should look for an exercise group, perhaps led by a physical therapist or by someone who can teach them the best way to begin and continue with aerobic exercising.

"It must become a way of life. It takes at least six weeks to see even very small physiological changes. It won't happen overnight and it must be continued."

Decisions about exercise are difficult to make, especially for those with more debilitating cases of FM. My support group included men and women whose conditions varied from mild to severe. There were several who made it a point to walk a mile or more a day, regardless of their pain. For those who could not walk a block without severe pain, it was hard to hear these people speak of "walking through the pain." And yet it seems that the authors of the medical journal articles and those who do manage to exercise or carry on their lifestyles with little change fault those who do not enjoy such freedom of movement.

It's almost as if they are saying, "If you would just make the effort, you, too, could do what you once were able to." That is a fallacy that is hard to fight, since it is often expressed in a skeptical tone of voice or with the lifting of an eyebrow. And that belief can add to your guilt and low self-esteem, which in turn can depress you even more.

So what do you do? Struggle to exercise or keep your activity level at an unrealistically high stage, allowing yourself to sink into a continuous cycle of pain, depression, and anger? You can't do that. Negative thoughts only contribute to the pain and poor sleep. You don't need to add to your guilt level. (If you are like many of us, you already have a lot of guilt about not being able to do what you once could without adding anymore to it.)

You must find the right level for yourself. Regardless of what you have done or felt in the past, at this point only look forward. You know how much FM affects your body, mind, and spirit, which in turn affects your lifestyle and those around you. Working with your doctor and physical therapist or with a group, determine what changes you must make. Start your exercise plan. How much can you do? At what level should you begin? What is best for your condition?

Follow your doctor's guidance. If you begin an exercise program supervised by your doctor and physical therapist, don't stop or decrease your program when released from the physical therapist's care. No one can afford to continue to visit the therapist forever. After you stop going, you must continue your exercises. Try to find an exercise group that is compatible with your program, or organize one with others who have FM. Those who have company in their exercise are more likely to continue the program and accomplish their goals. You also might look into the aquatic programs offered by the YMCA or YWCA. Just make sure they know something about fibromyalgia or you might do more harm than good even with the leader's best intentions in the world. Check into the Arthritis Foundation's People With Arthritis Can Exercise (PACE), since it is designed with those who have arthritis in mind.

If you can't find anyone, try it on your own. It's hard, but it can be done. If you are determined enough, you will stick with it—after all, it's to your advantage, it's your body and no one else can do it for you. All of the passive treatments and exercises in the world won't help if you won't work on your own. Deconditioning is unhealthy enough for a person with no health problems but because of the state of our bodies with FM, becoming or staying completely inactive only makes it worse. No one is asking you to run in a marathon (if they do—say no) or to climb Mt. Everest (will they carry you up and back down again?). You must have the desire to want to do as much as you can and only you can then carry out that desire. Just remember—be realistic in your goals and go after them slowly and in gradual increments. Then maintain them.

EIGHT

Cognitive Behavioral Training

The next step of the treatment process is that of cognitive behavioral training. Although we have divided up the various elements of treatment for FM, many of them really cross over into other areas. This section on cognitive behavioral training includes a few methods—like self-hypnosis and biofeedback—that used to be considered *alternative medicine*, and were not even considered a few years ago. But "the times, they are a' changing" and what was once thought quackery, at least in terms of Western medicine, is recognized as beneficial.

Once you start looking around, you will find that there are many resources, both professional and nonprofessional, that you can tap into to help make your life easier and more comfortable. Depending on the level of intensity of your fibromyalgia, you may or may not need all of the following resources.

Psychiatrists, Psychologists, & Therapists

Many people with FM want nothing to do with these care givers, but you should try to keep an open mind about them because there are many ways they can help. Both psychiatrists and psychologists deal with psychological problems, yet only psychiatrists are physicians, qualified to prescribe and regulate the antidepressants that are used to treat FM and depression. Unfortunately, depression can and often does, become part of living with a chronic pain condition. This is nothing to be ashamed of, although too many people refuse to acknowledge it and seek help. Obtaining a psychiatrist's help is important under these circumstances.

Many psychologists specialize in helping people to live with chronic pain and do much more than just one-on-one counseling. Many of them facilitate group therapy sessions, and teach stress management using self-hypnosis, relaxation therapy, and biofeedback.

Because traditional medicine provides little relief from chronic pain, many people feel the need to explore nontraditional methods to achieve relief. Within the last 20 years, some of these methods have gained wide acceptance. There are a number of alternative methods of treatment that directly involve the person seeking help. The methods are generally taught to the client, who then continues the practices or procedures independently. Usually, although not always, it is a psychologist who does the teaching. I learned my early relaxation therapy from a nurse practitioner who had a Ph.D. in counseling and pain control, but it was a psychologist who taught me the principles of biofeedback.

With the more common drug and physical therapy methods of treatment positively impacting only about 30 percent of those with FM, and even then only on a short-term basis, doctors are turning to methods used to alleviate other chronic pain conditions. The term cognitive behavior therapy is applied to the combined efforts of medical, psychological, social work, physiotherapy, occupational therapy, and nursing interventions. The therapy is directed toward self-management of the chronic pain. Besides common aerobic exercise, physiotherapy, and drug management, the individual is taught relaxation therapy, imagery, stress management, and avoidance of social stressors, communication skills, and an understanding of psychological reactions to stress and pain.

Programs may vary slightly from one clinic to another. The primary goal is to involve the patient in the management of his or her own condition. Some FM patients using behavior therapy show improvement not only in pain perception but also in psychological distress. This distress is often attributed to the feeling of lost control over symptoms as well as life in general.

This field of treatment offers significant possibilities for the individual, and many rheumatology clinics are advocating it for their patients' relief.

Cognitive Behavior Techniques

Stress management

Relaxation therapy

Self-hypnosis

Guided imagery

Biofeedback

The concept of conscious mental control can be traced to ancient times (even though the modern world is credited with developing the technology that makes biofeedback possible). These techniques all focus on the tie between the mind and the body and allow the individual to exercise a measure of control over both.

Just as with drugs, physical therapy, and manipulation, some of the treatments work for one person, while they may not relieve another person's pain or fatigue. It is also true that these methods work in cycles. They may deliver results for a while, but then the effectiveness wears off. If you try them again later, you may once more experience a period of positive results. Some techniques may have an initial high cost, while others may involve no more than the price of a book or tape and a time investment. You must decide if you are willing and able to make whatever investment is required.

Behavior modification is often used in the treatment of chronic pain because there is so little else that is effective. Pain is the body's natural warning indicator, informing the mind of tissue damage or illness. Acute pain follows an injury, operation, or illness that causes damage to the body, but it is short term. Once the underlying cause is removed or healed, the pain disappears.

This is not the case with chronic pain. Fibromyalgia has become one of the best examples of chronic pain syndrome. No underlying cause can be found, yet the pain is severe and lasts for months or years.

One theory on fibromyalgia is based on the hypothesis that something has shifted the central nervous system into "high gear," causing those with FM to have a very low pain threshold and tolerance. It takes less stimulus to feel pain, and there is generally more difficulty handling it for any length of time.

It is helpful to accept and understand that there is an actual link between the mind and the body. It is commonly accepted that stress can be a prelude to high blood pressure, a heart attack, or asthma. Stress also plays a role in FM. Alternative treatments recognize the effects of stress, regardless of how those effects surface in organic changes.

Stress commonly causes people to "tense up," contracting their muscles into hard knots. This comes from the body's instinctive "fight or flight" reaction to danger. In a healthy body, the accompanying flow of adrenalin leads to muscle contractions. Because muscle spasms are already a troublesome factor in fibromyalgia, this stress response only compounds the problem.

Many times we react to the pain in our bodies by tensing up even more, unconsciously adding to the problem. In order to obtain relief from this cycle of muscle pain, you need to learn to relax the muscles involved. Learning to do this requires a conscious effort, because most people are not familiar with the individual muscle groups. One method of relaxation teaches people to clench each group of muscles and then relax them so that they are aware of the difference. An example would be clenching the jaw and holding it that way for 2 to 4 seconds before releasing the tension. The difference between muscle contraction and relaxation then becomes quite obvious. Unfortunately, for many people with fibromyalgia, this process only makes the pain worse, so it may not be the best relaxation method for everyone.

A first step towards stress management and relaxation for those with FM should begin with proper breathing. Too often we don't breathe deeply enough because of pain or fatigue or some other reason. When someone is in pain, they have a tendency to hold their breath, waiting for the pain to pass. According to Katherine McCoy, "The diaphragm, which is the major breathing muscle, is often compromised in FM patients to such an extreme degree that accessory muscles take over, resulting in upper quadrant pain." She goes on in her handout "Teaching patients proper inhalation through the nose and exhalation through the mouth and nose is important for pain reduction and exercise endurance."

Many times at night when I am having trouble relaxing and getting to sleep, I will consciously control my breathing. In through the nose, pause and slowly let it out through my mouth. I will also breathe from my stomach, not my chest. I know that a lot of people don't think about how they breathe, but if you stop for a minute to focus on it, you will realize that you are breathing too shallow. One of the things I see over and over, in my research, is that there is inadequate oxygen flow, to the brain, through the muscles. One recent study found that individuals with FM had a significant drop in their overnight arterial oxygen levels as measured by the monitor that is placed on one of the fingers. So take the time to be aware of your breathing and work on it.

Relaxation Therapy

If you choose to attempt relaxation techniques on your own, Dr. Herbert Benson at Harvard Medical School explains how to do so in his book, *The Relaxation Response*. However, Dr. Benson recommends that anyone using the book for specific health purposes should do so under a doctor's supervision.

The basic concept of relaxation therapy is simple. It is based on bringing the body to a state of complete relaxation by emptying the mind of all worries and pressures. The general method for achieving this is to go into a room by yourself with no distractions (such as a stereo or television). Make yourself as comfortable as possible in a chair or recliner; some people lie on a bed or a couch. But when you use the technique during the day, you don't want to fall asleep; you should just work toward becoming gently relaxed.

Close your eyes and let all of your thoughts just float out of your head. It is not always easy the first few times you try it, but you don't want to aggressively push your thoughts away. If thoughts keep intruding, just imagine your self gently releasing them so that they drift out of your mind.

The first time you attempt relaxation, you may not be able to tell exactly which muscles are clenched or contracted. This is a good time to find them by consciously tightening and then loosening them. Begin with the muscles at the top of your head, then in your forehead, and on to your eyes. Feel the muscles in your jaws, let them relax so that your mouth feels almost ready to drop open.

Progress to your neck and shoulders and down your arms to your hands and fingers; move all the way down your back. Then come back to your chest and stomach. Move down to your thighs and then down your legs to your toes. There is no rush, no need to hurry. Take your time, so that you feel each group of muscles relax.

At this point you will experience a sense of warmth and lightness throughout your body that almost feels like you are floating. If you choose, you can use imagery, traveling in your mind to some restful place. It can be anywhere that you would enjoy being—a warm beach, an open meadow, or even a soft, pillowy cloud. Stay there until you feel a sense of peace and rest emanating throughout your mind and body. Then slowly and gently return to the room again. Again, don't let outside thoughts intrude on your return; let them just drift away.

The entire experience should take about 10 to 15 minutes, and when you are finished, that sense of peace should follow you back into your regular day. It won't relieve all of your problems or pain, but it will ease some of the muscle spasms that increase your pain.

Self-hypnosis

From the very simple exercise above, such methods as self-hypnosis and biofeedback have been developed. Many people don't believe that they can be hypnotized, and even though some can't be, most find they make good subjects. If possible, have your doctor refer you to someone who can teach you the basics and determine the best method of hypnosis for you.

Costs vary depending on your area and the availability of hypnotists offering their services. Some university psychology departments offer clinics with reduced rates or provide services through a rehabilitation program. Check with the colleges or universities in your area. As with other alternative methods of treatment, sometimes hypnosis only works for a while before becoming ineffective. For example, some people who use self-hypnosis switch to a different set of tapes when they feel that the old set is losing its effectiveness.

Biofeedback

With the help of personal computers and special software, biofeedback has become much less expensive and is now a feasible option for many more people. The concepts of biofeedback and relaxation are very similar.

In biofeedback the amount of tension is measured with small electrodes that can be placed on your finger and forehead, or on any set of muscles that you need to learn to relax. The equipment used in the early 1970s required the use of big complex machines, which were attached to electrodes that measured the temperature in the muscles. You can learn to lower your temperature and relax your muscles by concentrating. The tension in the forehead registers as a series of beep tones that are very fast when you are first hooked up but that slow as you learn to ease the muscles. Regardless of the specific setup a particular psychologist uses, the goal is the same: to teach people that they have control over their bodies and that they can learn to relax tense muscles.

Stress Management

Stress management may be incorporated into the techniques for relaxation therapy, self-hypnosis, and feedback. But it is also much more. Stress management is designed to help you cope better with overall life including your work, health, familial situation, and more.

The theory is to identify what stressors are present in your life and how they affect you and impact your pain. The next step is to develop ways to relieve some of them and to find methods of coping and living with others. Stress is defined as a mental or physical tension or strain. You already know about the effects of stress on your body, specifically on your muscles. Having fibromyalgia is stressful. The condition is inherently full of uncertainty. One day you may be able to do your job and get a great deal accomplished; another day, you can only do the minimum required to take care of yourself. This type of uncertainty is frustrating for you, and for your family and friends, who see no reason why the days turn out the way they do.

Pain itself is stressful. When it goes on without relief day after day, you begin to lose the ability to cope with it. Many of the emotions that accompany FM only add to your stress.

Sometimes everything in your life seems determined to become difficult simultaneously. Just when your FM is flaring up and you have less energy and more pain, a major crisis occurs at work. Just when you think you have that under control, the washing machine breaks down and floods the laundry room while your budget is stretched so tight it's screaming. Or the transmission goes out on the car that you were trying to make last another year.

Everyone's life has times like this. No one's life always runs smoothly. So why should we expect it to change for us just because we now have fibromyalgia? We can't. What we can do is try to be prepared for such times and use a psychologist's help to get through them. It might even be worth the time and money to join a group therapy session for a while. If possible, join one that is composed of others who are experiencing chronic pain.

It is difficult for many people to tell their deepest fears and emotions to a relative stranger, even if that person is qualified to help them regain some stability. The idea of sharing one's deepest feelings with a room full of

strangers seems even worse to most people. Participating in a group therapy session may seem impossible for you but, particularly if it is a chronic pain
group, it may be the best thing you have ever done for yourself.

You will find out that you are not alone in being afraid, angry, and in feeling less confidence in yourself. You will also discover how others cope with
their emotions. Usually there are a set number of weekly sessions, such as a
six-week period of sessions, and you will save money on the cost of counseling by meeting in a group.

By participating in a group, you build up a support system for yourself.
This is particularly important if you have no close family; you need to find others who will listen when you need to talk and who understand what you're
going through when you need to cry. This is an informal arrangement and
one that should be flexible enough to meet your personal needs.

One study on fibromyalgia found that people with FM seem to experience more stress from the hassles of daily living than they do from larger
stressors like death or divorce. If this theory is proved, it indicates that we
need to look for ways to help us cope with everyday hassles, which can
include anything from the crime rate to traffic to running errands.

*Accept your illness. Educate yourself concerning it. Pace your
living. Keep busy and help others with the syndrome. Laugh a lot
and pray. —Evonne*

Stress management generally covers everything from relaxation therapy
to time management. The idea is to give you ways that will help you get
through your life without feeling like you want to run screaming. You may still
want to run away, but instead, you can choose to deal with the most important matters first, which you have learned to prioritize. And you may choose
to never deal with those problems that are lowest in priority. You must learn
to recognize the stressors in your life, both positive and negative, then determine the best way to deal with them. Using your time and energy effectively
allows you to accomplish more, but old habits are hard to break. If you feel
better, you may slip back into pushing yourself more than you should.
Sometimes, you will succeed without triggering any problems and sometimes you won't. Remember, you are in this race for the long run.

Self-talk

Self-talk is learning to communicate with and control the inner voice within us
all. Self-talk affects your well-being and self-image. As conscious beings, this
constant communication going on in our minds is closely tied to our view of
ourselves, even though most people aren't really aware of that communication or of its implications.

The voice within you is based on all of the things you were told as a child. If you grew up in a home full of love and caring, and you were supported in your endeavors as you became an adult, your inner voice reflects that loving confidence. But if your home was not supportive, or dysfunctional, you probably need to re-evaluate what that inner voice is telling you.

If you grew up with a very low sense of self-worth, your inner voice may be adding to that feeling because your FM prevents you from fulfilling your roles in life, whether as spouse, parent, money-earner, or all of these. If pain prevents you from working, your feelings of uselessness may be confirmed.

A psychologist can help you cope with your negative self-talk, or you may simply need to become aware of what you are saying to yourself. You need to be able to take an honest look at yourself to determine whether you are truly doing the best that you can under the circumstances. If you are doing your best, then you need to accept that. If you find that you're not doing as well as you could, then you might find that a program of positive self-talk will help.

Motivational speakers advise their listeners to repeat a positive sentence or two a number of times during the day. It can't hurt to develop one or two of your own. Good examples might be, "I know that I am doing the best I can," or "I will make two contacts with someone outside my home today."

Of course, you must be honest with yourself. You must really be trying to do the best you can. Self-honesty is sometimes difficult and painful. But it is always necessary. You must believe that you are wholeheartedly trying to live and work in spite of the pain. Only you can really know whether you are surrendering to the pain or living your life in spite of it.

Laughter

One last suggestion. Seek out laughter as medicine. Norman Cousins first wrote about the use of laughter as a form of medicine in his book, *Anatomy of an Illness*. He had a type of arthritis, and his doctors weren't very encouraging. Cousins decided to try laughter and was soon watching many of the old slapstick movies of Laurel and Hardy, the Marx Brothers, and the Three Stooges. He experienced a dramatic improvement.

For myself, I have found that such movies as *Foul Play*, *Seems like Old Times*, and *Beethoven* have me laughing out loud. Not only am I not aware

> *My tip is having power of prayer, faith in God and keeping a positive attitude. Being able to laugh at myself. I like to read a good mystery that takes my mind off my pain. Knowing what is happening to my body helps with the positive attitude.* —Jo Ann

of my pain during these movies, but I have also found that I feel better afterward. Laughter encourages the production of endorphins. In contrast, action movies, such as *Jurassic Park*, get the adrenaline flowing but leave me totally exhausted and unable to sleep for some time.

I don't know the scientific facts behind these findings, but I do find I prefer to laugh myself out of pain rather than be scared out of it.

NINE

Complementary or Alternative Treatment

Doctors of traditional Western medicine have long refused to accept anything that isn't covered in their medical school training and unfortunately, Western medicine has divided the mind from the body, believing every ill was either of the mind or of the body. Despite the controversy it can arouse, there has been a great deal of interest in the area of alternative or complementary medicine lately and many doctors have become much more willing to consider alternative or complementary methods of treatment in cases where traditional medicine has not been able to provide relief or a cure, usually in addition to their own prescriptions and recommendations.

Fibromyalgia is a prime example of why people turn to other methods. Despite the fact that we have made tremendous strides is medicine, there are still many conditions for which there is no effective treatment. According to an article in *Primary Care* (December 1997), fibromyalgia, along with back pain and arthritis, contributes the greatest clinical and economic burden to overall chronic pain statistics. And these patients are among the most frequent users of alternative therapies.

When people are desperate to get relief from pain, fatigue or other symptoms, they are also at their most vulnerable. At best, traditional treatments available for FM are effective in 30 to 50 percent of cases. In many of those cases, the relief is only partial. So individuals look for something, anything, that will make the pain go away and will give them back the life they've lost. The article above described the University of Maryland School of Medicine's Center for Complementary Medicine Research's progress towards investigating alternative methods of treatment. According to the article, the most commonly used therapies include acupuncture, homeopathy, manual/ manipulative therapies, and mind-body therapies. The study concludes that these therapies have the "greatest clinical potential for these three conditions (i.e., FM, back pain and arthritis)".

Anyone who considers some alternative or complementary treatments should read two books first. *Manifesto for a New Medicine, Your Guide to Healing Partnerships and the Wise Use of Alternative Therapies,* is by James S. Gordon, M.D., who is the first chairman of the Advisory Council of the National Institutes of Health's Office of Alternative Medicine and is also Clinical Professor of Psychiatry and Community and Family Medicine at the Georgetown University School of Medicine. He is also director of The Center for Mind-Body Medicine in Washington, D.C. The other book is *Dr. Rosenfeld's Guide to Alternative Medicine, What Works, What Doesn't and What's Right For You,* by Isadore Rosenfeld, M.D., who is Rossi Distinguished Professor of Clinical Medicine at New York Hospital/Cornell Medical Center and attending physician at both the New York Hospital and Memorial Sloan-Kettering Cancer Center. He also served for several years on the Practicing Physicians Advisory Council for the Secretary of Health and Human Services.

I believe that these two books will give you a good start on determining which form of alternative treatment you should consider and how you should go about evaluating the worth of such treatments. And you should consider any treatment carefully, whether traditional or alternative, because you may be much more sensitive to physical, medicinal, or supplemental therapies.

According to Gordon's book, a survey published in *The New England Journal of Medicine* (1993), showed that Americans were paying $10.3 billion out of their own pockets for alternative treatments in 1990 (the date the survey was taken) and they made 425 million visits to practitioners of alternative therapies. The majority of the time, they did not tell their regular doctor about that treatment. So people are open to alternative treatments. If you are going to be one of those who seek relief from the pain, fatigue, and poor sleep, of FM, make sure you are prepared to consider carefully the pros and cons of each treatment. (You will find that Dr. Rosenfeld is more skeptical of some forms of alternative treatment than Dr. Gordon, who actually practices some of them.)

Alternative health practitioners practice the concept of holistic medicine, the close functioning of mind and body. Some principles of holistic medicine are:

- The mind and body is a single entity
- Good health is a positive state, not merely the absence of disease; the patient must work to preserve it. Prevention, not treatment is the first priority
- The body can often heal itself without medication
- The whole patient is treated rather than any one organ or system

Alternative, complementary, and holistic therapies pay more attention to the impact of the total environment on bodily functions, delving more throughly into the kind of food you're eating, whether or not you smoke, how much you drink, how much you exercise, and how happy you are with your lot in life.

This book isn't designed to cover all of the elements of alternative medicine but it will provide you with brief explanations of those that might be helpful in dealing with fibromyalgia. The listing of such treatments is not intended to be an endorsement nor a promise of a cure. Many leading doctors on FM encourage patients to try alternative treatments. But do so very carefully and very slowly. Change only one thing at a time and give the treatment a chance to work. If you see no improvement within a month to six weeks, it's probably not going to work. If you see any strong side effects, stop the new treatment immediately.

There are other doctors who will never be open to the idea of anything but Western traditional medicine, but probably most of them have chosen not to recognize fibromyalgia in the first place, and, in the second, they will tell you "You'll just have to learn to live with it." Gee, I hate that phrase.

As mentioned earlier, there will be some crossover between this chapter and the previous two as many of the "body-mind" treatments are becoming the standard, especially in the treatment of chronic pain. We have already touched on most of those, relaxation therapy, self-hypnosis, biofeedback, and guided imagery.

Physical Intervention

Physical treatments like osteopathic or chiropractic manipulation and massage therapy have shown some of the best results. These are most effective in conjunction with other modalities. Not everyone with FM is able to tolerate such treatment, and everyone should approach it cautiously. I went to an osteopathic physician in Fort Worth, Texas, who specialized in manipulation and who provided me with relief for a number of years before it became clear that I could not afford to continue the cost of treatments . . . not to mention losing most of the benefits of the treatment during the long drive home.

Osteopathic medicine believes the musculoskeletal system is intimately related to overall wellness. Manipulation of the musculoskeletal system aids the body by promoting healing and by maintaining a high level of health, through increased blood flow, nerve conduction, and lymph flow to and from affected areas. This is accomplished by moving the skeleton and muscles in the body to remove any muscular or bone impingement that may be affecting the blood flow, nerve conduction, or lymph flow.

Chiropractic philosophy believes the nervous system connects to every part of the body by way of the brain, spinal cord, and nerves and controls all of the bodily functions. Chiropractics is based on the belief that misaligned vertebrae can "choke" the nervous system, resulting in poorer health. By removing those misalignments, chiropractics can help the body heal itself without the use of drugs and surgery.

Massage therapy is often used in conjunction with physical therapy. Massage therapists are often state licensed. I have an independent massage therapist who has been working with me for the last 2 years and I get a lot of relief from him. Ask potential massage therapists if they are aware of fibromyalgia or have ever treated anyone with it. Like many other health care providers, massage therapists have areas of special interest. When I first started going to Justin, he didn't know much about FM but he worked with me, starting gently, learning where my worst spots were, and what depth massage I could handle. He has made a major difference in my life and I just wish I could see him more often.

Finding a massage therapist who is knowledgeable about FM also applies to osteopaths, physicians, and chiropractors. Some people have had very good luck with seeing one or the other but, for some, these hands-on treatments can make their pain much worse. When you first begin a manipulative therapy, you may experience more pain; you should then gradually improve and each treatment should provide you with some relief. If, however, you have a significant increase in your pain, your practitioner is working too hard or too deep, or this mode of treatment may not be right for you. Just as in the exercise program, go slowly and gradually.

Acupuncture is increasingly recognized as an effective form of treatment and a complement to traditional medicine. This Chinese procedure has been practiced for more than 2,500 years throughout Asia and Europe and involves the insertion of tiny threadlike needles along the meridian lines of the body. The general theory of acupuncture is based on the premise that there are patterns of energy flow called Qi (pronounced chee) throughout the body that are essential for optimal health. Studies indicate that acupuncture influences the central and peripheral nervous system. (Needles are used once and discarded, hence, the risk of disease transmission is remote.)

According to a Consensus Statement on Acupuncture recently released by the National Institutes of Health, Americans spent more than $500 million per year and made between 9 and 12 million patient visits for acupuncture in 1993. The consensus statement commended the ongoing increase in improved training and called for more uniform licensing, certification, and accreditation of acupuncturists, which would "help the public identify qualified acupuncture practitioners and to have more assurance in quality of service." At the time of the Consensus Statement, 34 states licensed or otherwise regulated the practice of acupuncture by nonphysicians with establishedtraining standards for certification.

Adverse side effects are mild and the panel recommends that patients be fully informed of treatment options, risks, and safety practices. They also recommend that federal and state health insurance programs including

I have two factors that give me hope:
1. *Keeping informed about research development and hints from other sufferers (via Internet).*
2. *Letting go of grief and resentment.*

—Joan

Medicare, Medicaid, and other third-party payers, expand their coverage to include appropriate acupuncture treatments.

Acupuncture has been found to be very helpful in chronic pain conditions, pain with cancer, post surgery pain and in relieving nausea in pregnancy.

Acupressure is adapted from acupuncture. No needles are used; pressure is applied to the same meridian points as in acupuncture.

Tai Chi is another form of therapy. Tai Chi is sometimes called "Chinese shadow boxing." Though it is considered a form of martial art, it involves smooth, flowing movements that resolve into forms inspired by animals. An article in *Harvard Women's Health Watch* (November 1996), called it a "state of constant, unified motion in which the goal is not to isolate or concentrate on individual parts of the body but to be aware of the hands, head, body, and limbs in concert. The leg serves as a base and the waist as the axis of movement."

The entire Tai Chi routine can be done in as few as 10 minutes. It provides a very low impact aerobic workout and requires only about 5 square feet of space. The only dress requirements are loose clothing and flat shoes with soles that won't stick to the floor.

The Chinese use Tai Chi to treat a number of health conditions including circulation and nervous system disorders, addictions, muscle injuries, arthritis, and even asthma. It has been used for stress reduction as well.

Diet

Even though research has not found diet to influence FM, some people experience sensitivity to certain foods, even some they have never had problems with before. Maintaining a normal, healthy diet is essential in dealing with the overall debility of FM.

One problem that you may encounter is an increase in weight, which may be due to your medication or decreased activity. Many factors may play a part in increased weight. The poor self-image that comes with weight gain is also a source of additional emotional stress. Although it has not been confirmed by specific studies on fibromyalgia, it is believed that people with chronic illnesses often have lower metabolism rates. It may be wise to assume that your metabolism rate is slow, especially when looking at nutritional supplements, herbs, and other nonstandard food items. Because FM has such a complex and individual impact on each person, he or she may not only process medications faster or slower than what is considered the "norm," but that may also change over time. One day the dosage may essentially knock him or her out, making that person sleepy, and, on another day, the same dosage may seem to have no impact at all. Therefore, the speaker at one FM conference recommended changing only one part of his or her treatment at a time. For example, if the person has decided to try a new supplement, he or she should take that one for a period of time, not introduce two or three new ones at the same time. This can also be applied to adding new foods to your diet.

It is not unusual for people to eat more as a way of dealing with the stress of chronic illness. Not only is extra weight unhealthy, it puts more stress on the muscles and bones.

It isn't easy to make dietary changes, especially if you live alone and have less energy to cook healthful meals. The Arthritis Foundation recommends the following general guidelines for anyone with arthritis, and these also apply for those with fibromyalgia:

- Eat a variety of foods
- Maintain your ideal weight
- Avoid too much fat and cholesterol
- Avoid too much sugar
- Eat foods with enough starch and fiber
- Avoid too much sodium
- Drink alcohol in moderation

The Arthritis Foundation also warns about combining alcohol and other substances. Stomach problems are more likely to occur if you drink alcohol while taking nonsteroidal anti-inflammatory drugs or aspirin. Consuming large amounts of alcohol with an acetaminophen can damage the liver. The potential for serious side effects with prescription drugs is even greater.

Other doctors recommend avoiding caffeine, found in sodas, coffee, tea, and chocolate, because it is a stimulant.

Initial studies on the efficacy of herbal remedies and supplements like Melatonin, DHEA, malic acid or magnesium have been inconclusive. If you are considering taking these or any supplements, follow the recommendations given by Dr. Daniel Clauw at Washington:

- Change or take only one supplement or herbal remedy at a time
- Take the minimum amount recommended or less
- Take it for one month to six weeks; if there is no improvement, discontinue it
- If you believe you are feeling better because of a particular treatment, stop taking it for a few days. You should feel worse.

• Keep a careful record of what you are taking, the amount, and any side effects you experience

If you experience unexpected side effects, discontinue taking any supplement or herb immediately. I would also recommend that you be somewhat skeptical whenever you hear about the miraculous properties of a particular supplement or new treatment. Check it out as thoroughly as possible. Consult your physician before taking any supplement or before making any major dietary changes.

Do your homework on whatever you choose to try, then conduct your own study, keeping a record of how you feel, how well you slept and other pertinent information. Don't invest a lot of money in any of these alternative treatments until you are sure they work.

TEN

Adjusting to the Changes

Fibromyalgia takes you into a new world. Whether you have been there for years as you searched for a diagnosis and effective treatment or have just arrived as a newcomer whose symptoms appeared only recently, this world is probably one filled with anger, fear, frustration, and pain.

If you are like many people who have been misdiagnosed and struggled through treatment after treatment that brought no relief for your pain and fatigue, you're probably feeling more than a little anger. You have a right to be angry. With today's advanced technology, we have come to believe that doctors know so much that almost anything should be easily diagnosed and treated. There should be a pill for whatever ails you. But all the shots in the world won't change the fact that medical researchers still have a long way to go before they can recognize and treat every medical condition that arises.

Fibromyalgia is only one of hundreds, if not thousands, of medical conditions that doctors have had a hard time diagnosing and treating. But does that fact make your life with FM any less painful or easier to live with? No. And you shouldn't expect it to. No matter how many times you are told to be thankful that FM is not life-threatening or crippling, it doesn't make facing each day of pain and fatigue any easier. Your life has changed. You are the one dealing with the pain and fatigue. You are the one who must live with that 24 hours a day, every day.

So what can you do? Withdraw into your secluded world of FM? It may be tempting. But you have a choice about how you are going to live with FM now that it is a part of your life. One step in beginning to cope might include adopting "The Serenity Prayer."

When I resigned from my last full-time job, I wrote "Serenity, Courage, and Wisdom" on a small blackboard and hung it where I could see it every day. As you will learn, each day is a new day, and even though you may believe you have conquered the anger, fear, and depression, the battle is

---

The Serenity Prayer

God grant me the serenity to accept the things I cannot change, the courage to change the things I can, and the wisdom to know the difference

---

never really won. Some days are easier and some days are harder, but everyone must be faced—faced and lived.

It would be so easy to resign from life, to sit down in a chair and refuse to do things that you know will cause pain. It may be easy, but it's not something that you can do. You cannot give up. There is something within all of us that does not want to acknowledge giving up. Regardless of the pain and fatigue, you need to spend the rest of your life living life to the fullest, whatever that may be for you. Whether it's the bookkeeping system or the parking arrangements, people tend to resist change.

And now you are faced with changes that you don't want to make, changes that are being forced on you. As an adult you may have never taken a daytime nap, even though you walked around in a fog of fatigue, because you were determined to keep going. Now you may find that a short nap, or at least a brief rest on the couch, will enable you to put in more work time in the long run. Look at these changes as a positive way of making life easier for yourself.

Coping With the Anger

When I realized that I would no longer be able to get and keep a full-time job, it made me angry. It also scared me. After all, I'm single and must support myself. It had been hard to manage financially when I was working full time. What was going to happen now that I couldn't keep that full time job? The thought frightened me, and the fear and anger fed on each other in a never-ending cycle. For over 4 months, I walked around wanting to put my fist through a wall.

There are productive ways of expressing anger. It is much healthier to express the anger than to keep it buried. You must learn and incorporate safe

The aftereffects of a low-impact aerobic class will enhance your everyday life. I feel strongly that the exercise must be structured so that your heart rate goes up and you experience the endorphin high . . . and a little sweat! The effort is worth it as your body becomes more flexible and you increase your endurance. —Susan

Before bedtime I list things I hope to accomplish the next day, realistically estimating time needed. Having definite goals helps me face each day positively. —Jeanne

ways of expressing anger so it doesn't smolder and eventually erupt. Besides the physical cost of holding in anger and the stress of controlling an outburst, you may also damage a lot of relationships.

Webster's New World Dictionary defines *coping* as "to fight or contend with successfully." Presumably, you are learning as much as you can about fibromyalgia, which is an indication that you have chosen to cope. Coping with any chronic illness is a constant fight. It takes a conscious effort. Living with fibromyalgia is not easy. Your course of action must be deliberate and planned. Even though there is always the element of uncertainty, you must develop alternative strategies that you can use to meet that uncertainty.

The first step is making the decision that you are going to live as well as you can. The second step is forming a partnership with your health team and establishing goals. And the last step is facing the emotional impact of the changes you are experiencing.

Handling the Changes In Your Life

Whether you have already confronted the changes in your life or they are just now appearing, you need to recognize that everyone fears and resists change. Even when you are excited and looking forward to a change, there is still an element of reluctance. How often have you heard someone say, "But that's the way it's always been done" when asked to make a change?

By expressing your anger, you have recognized your legitimate right to get angry. You must also decide that you aren't going to let anger become a disruptive force in your life. Although you have the right to feel angry, you don't have the right to make yourself and everyone else around you miserable. On the other hand, don't burden yourself with guilt about being angry in addition to the anger. Whether you express it by having a good crying spell or by throwing something, you must be able to let it go. Your fibromyalgia won't go away, nor will most of the circumstances change, but your ability to live with them will improve.

Will one expression of anger, one crying spell, or the satisfaction of seeing something break, end all of your anger? No. You will still get angry, particularly when there is something you want to do, but don't have the energy to do or cannot do because the pain is too bad.

Don't misunderstand; it isn't easy to learn how to vent your anger in an acceptable, healthy manner. Your first reaction may be to find a physical release for your anger. But when this sort of activity can severely aggravate your condition, this is not a realistic option. Don't be afraid to go to a psychologist or counselor who can help you find effective and appropriate ways to release and cope with your anger. You may need some help.

Sleeplessness

Anyone with chronic pain may have many sleepless nights. You may have them regularly at certain times, or you may not have any trouble at all. Sometimes, you will be so tired that you can't imagine why you can't sleep.

How do you manage to get through those long hours? Try reading. Reading will help you get through the night, and it can also help you cope with the pain by diverting your mind.

Feeling All Alone

Loneliness may plague you more often these days. You may feel removed from everyone else. No one understands just how tired you are or just how much the pain hurts. You feel isolated and alone. But there are others who do understand. They may have fibromyalgia or some other chronic illness. Realizing that you are not alone is one of the most important things you can do. You may find others within a support group or even in your day-to-day contacts who are struggling with the same experience. However you find them, these friendships will help to ease the loneliness.

Dating and Personal Relationships

Friendships are essential to your mental stability—time spent with your family, children, spouses, parents, or brothers and sisters—can still be socially fulfilling, but you also need the joy that comes from social interaction with others.

Make the effort to get out of the house periodically even if it's only to meet a friend for lunch or a glass of tea. The change of scenery will renew your spirit, and the outing doesn't have to be expensive.

If you are single when is there time and energy for a social life that includes dating? Although it may seem as if there are neither, it is important that you at least make the effort. You may not feel up to a night on the town. If you have always loved to dance, it can be disappointing to have to sit out every song. And late nights may be totally out of the question now. How do you continue to meet new people? What do you do about dating?

Try to get out at least once a week or every two weeks. If you don't have the energy for dancing, go to hear a special performer. Be well rested before you go, and try to make sure you have a comfortable chair to sit on. You may not be able to sit through an entire show, but it is a wonderful way to enjoy the music without feeling left out.

What do you say about your fibromyalgia to a potential romantic interest you've just met? Although you may want to get your health condition out in the open immediately, you don't want to pour it all out on a first meeting or even a first date. Unless he or she has asked you to go hiking, there's no need to bring it up until you both decide that you want to see more of each other. You will need to find the line between using a lot of "I can'ts" and refusing to discuss the reasons behind your limits.

If your relationship grows, some of your limitations will naturally become apparent. If the person you're dating is very involved in physical activities, it is best to be realistic and honest about your condition from the beginning of the relationship. If continuing intense physical activity means more to your partner than continuing a relationship, let go of the relationship before you become involved in a painful situation. For anyone who is more interested in getting to know you better, your fibromyalgia won't matter—it will just be a part of you.

Sometimes you may feel as if you just don't have the time or energy to even consider dating. That is all right, too. After all, you must set your priorities. But don't use FM as an excuse to hide from a relationship. Sometimes people feel that FM has stripped them of their attractiveness and they choose not to be open to meeting someone new. Concentrate on the aspects of your attitude and appearance that you can do something about. Don't agonize over those you can't change. You have to make the effort to meet new people, regardless of the complications involved. No one will come knocking on your door to improve your social life. Remember, you need a balance of work, play, and rest.

Sex and Intimacy

"My husband has stopped approaching me about sex. I'm just too tired after a day's work and then coming home to cook supper and clean up."

"It hurts for someone to even bump into me, much less some of the affectionate hugs and touching my wife and I used to enjoy."

Many times couples find it very frustrating to handle a chronic pain condition and still maintain intimacy in their relationship. Even today, many healthcare providers are hesitant to address the topic unless their patient brings it up. Anyone who considers asking a doctor about sex and intimacy may rightfully worry that such questions will be dismissed. If you are lucky, however, your doctor may be very willing to talk it over with you and is only waiting for you to ask.

No one can question the importance of sex and intimacy in marriage, yet what can you do when sometimes even a light touch brings pain? How do couples who have always had a very physically intimate relationship cope with the reality that such touching may cause one partner pain? It wouldn't take too many sessions of painful contact before a couple becomes too afraid to touch. And if just touching causes pain, what about lovemaking? Does a couple have to give up their sexual relationship completely? If the marriage wasn't overly strained before, this could overburden it completely.

Even if your doctor is reluctant to discuss it, there are sources of help and information. Your occupational therapist can provide plenty of information on the physical aspect of relationships. An occupational therapist can give guidance in choosing times and possible positions that would put less strain on affected muscles.

If you are uncomfortable speaking with someone directly, there are a number of books that can be helpful. Check out the health and self-help section at your bookstore or local library. The Arthritis Foundation publishes a booklet on sex and arthritis titled "Living and Loving," which also offers some helpful hints.

Although spontaneous intimacy is usually preferred, carefully planned moments of intimacy can be just as exciting if approached with the right frame of mind. After all, if you have taken the time to be well rested, the experience is more likely to be enjoyable than if you have to cut it short because you ran out of energy.

Depending on the location of your pain, different positions can prevent lovemaking from adding to the stress or strain on those muscles involved. Together you and your partner can explore the possibilities. A warm bath to ease your muscles and a dose of analgesics taken in time to provide some relief might help to make intimate moments as painless as possible.

Although many couples make love in the evening, those with FM enjoy no benefits by doing so. By evening, their fatigue has generally caught up with them, especially if they have been active. Mornings may not be the best time either because you may awaken stiff and in pain. For many, the best time to make love is late morning or early afternoon. Experiment. Don't be afraid to be innovative because it won't damage your body or accelerate your condition. After all, you want to nurture your marriage as much as you can under the circumstances.

If you are single and not involved in a long-term relationship, it will be more difficult. A great deal will depend on the person you date and his or her attitude toward fibromyalgia. At a time when sexual issues are discussed more openly, you should be able to talk about this with your partner. Some choose not to become involved in a sexual relationship so they won't have to worry about confronting these issues. Be careful about taking this approach. A supportive partner can bring so much joy to your life that you will be glad you didn't avoid an intimate relationship. And as Connie A. O'Reilly, a clinical psychologist on the Oregon Fibromyalgia Team, recommends, consider sex

> *Whenever I go into a restaurant, I take my cushion and sweater in with me that I keep in my car at all times. A straight, hard chair is like a torture chamber and getting cold is just as bad!* —Barbara

for one. She states in an article on the web site *myalgia.com* "Masturbation is a normal, healthy, and satisfying form of sexual activity. You can learn a lot about your body, what feels good, and what leads you to orgasm. It's time to bring masturbation out of the bathroom and the back bedroom, and into healthy sexual relationships, with a partner or by yourself."

Facing the Fear

Fear is very much a part of learning to live with a chronic condition. It may not be unusual to experience fear that is so strong it causes extreme panic. In the beginning, it may be difficult to believe that there is a limit to such fear and panic. You may fear the consequences of being unable to support yourself or of being incapable of meeting your commitments as a member of a family. The fear may also be rooted in knowing that constant pain is going to be a part of your life from now on.

Just as it is acceptable to feel anger, it is acceptable to feel fear. But you cannot let it completely command your life. You must exert some control, control that you may feel that you will never have again. Much of the fear expressed by those with FM is related to the loss of direction in their lives. There are some things you will never totally control again, and that is why it is so important that you make the effort to maintain whatever activities you can. Fear, particularly fear of change, is understandable and acceptable. But do not let it, and the other emotions you are experiencing, rule your life because they will only contribute to your symptoms.

FM's Effect on Your Self-Image

All of us carry around an image of ourselves inside of us. That image is not how others see us, but how we see ourselves; it shapes all our actions and thought. All of the things that make up the person that is you are encompassed in that self-view. Your physical shape and size, the color of your hair, your eyes, your skin, and the way you dress and carry yourself are all factors. The roles that you play help to determine that self-view in both positive and negative ways. If you have a successful career and a happy marriage and home life, you will generally see yourself in a positive light.

Deep within you, you have certain expectations of yourself—certain ideas that determine how you behave. If for some reason you are unable to meet the expectations that not only you have, but that others have for you, your self-view will be affected. Fibromyalgia can seriously affect that image. Perhaps you can no longer provide for your family financially and making love with your husband has become too painful or requires energy you don't have. Or perhaps it has become impossible for you to plan special outings for your family because you're never sure how you're going to feel. Or maybe you can no longer keep working at a job outside your home, which is necessary for your family's financial welfare. Being unable to fill these roles will inevitably have a detrimental effect on your self-image and happiness.

Fibromyalgia affects more women than men, and most of those women

> *I went through a five-week pain program. They offered many tips. I learned Tai Chi, biofeedback, relaxation techniques, and exercise (mild). The most helpful was massage therapy (total body). You would not believe all the trigger points I had. Pool and spa therapy was also helpful. I got relaxation audio tapes that help and also some self-hypnosis audio tapes. —Kim*

are in their 30s or mid- to late-40s. Many of these women came of age before the women's liberation movement began and were taught that they would be judged on the quality of their housekeeping. It may not have been an overt lesson, but it was one that they absorbed. Yet many women also work outside the home. Almost any woman's magazine contains articles that demonstrate that regardless of the post-1970s consciousness raising, women are still doing the housework. Most of us come home from a 40-hour-a-week job to another full time job of laundry, child care, errand-running, and housework.

This routine is enough to burn out a woman who has a normal energy level, so what can it do to someone who runs on half a tank, *if* she's lucky? Many of us still associate the image of self-worth with maintaining a house. And when we can't measure up to our own standards, even if our husbands or partners don't insist on such standards, our self-esteem is affected.

When you add the problem associated with keeping an outside job to the low self-image as a housekeeper, parent, and spouse, it becomes a grave situation. When you add that to the pain and frustration of a poorly understood, difficult medical condition, you are left with the potential for some serious psychological problems.

Because many of you have been told that your pain and fatigue is due to

> *I find walking every day, early in the morning, helps me become more flexible and helps me adjust my attitude when the pain becomes difficult to handle. —Janet*

psychological problems, you most likely don't want to hear that there may be such problems. But these psychological problems are caused by your pain, other symptoms, and the changes that are taking place in your world. And that is only natural. But how are you going to deal with them? You must realize that you are grieving and that you must work your way through the grief process, emerging with a more acceptable view of yourself.

The Grieving Process

Grieving occurs whenever we lose something or someone. We realize that it is natural to grieve over death and many also realize it is part of the divorce process. But is it all right to grieve because you have fibromyalgia? Haven't you lost something? Your health? This is usually more apparent to someone who has only recently developed the symptoms of FM.

It may have been only a few months ago that you took that great hiking vacation or that you spent all day shopping at the mall, ending the day with a special dinner for two. Now you find it difficult to stay at the grocery store long enough to get your usual supplies and just the thought of physical exertion leaves you out of breath.

Whether you were active or not, by the time you accept fibromyalgia, you have lost a lot. It may be the ability to work as long as necessary to put your ideas into actions. It may be horseback riding and owning your own horse or taking long hiking vacations. It may be climbing the corporate ladder to reach the position you always wanted. Whatever you have lost, recognize that legitimate loss has occurred in your life.

These losses are tied into the grief process. It is natural for you to experience grief. Do not feel guilty. It is all right to grieve.

Denial or isolation, anger, bargaining, depression, and acceptance are the stages of grief, and you must go through each of them. But you must not let yourself get stuck in the first four phases of grief.

It is a very personal journey, and no two people experience it the same way. Some people will move through the stages of grief in a different order, or even repeat them. Some will remain stranded in the first stage of shock,

---

Stages of Grieving

Denial/Isolation

Anger

Bargaining

Depression

Acceptance

while others may move rapidly through the first two stages of denial and anger to bargaining, or the abyss of depression. The length of time you spend in any of the stages isn't as important as the fact that you make your way through the process and come to terms with your loss.

Use common sense in facing your loss. And be gentle with yourself. After all, you didn't deliberately get fibromyalgia. You can be angry at the FM, but you don't have any reason to be angry at yourself.

Human nature cushions us somewhat from the first blow of grief by staging a period of denial during which we may try to isolate ourselves from the truth.

Usually, in any loss there is a period of shock, almost to the point of numbness, which is the body's way of easing you through the change. It is natural to initially deny your FM, to believe that it really won't have an impact on you. That isn't much of a problem to someone who needs only a mild analgesic to control their pain, but it is unrealistic for those who must alter their lifestyles. This denial is usually accompanied by a sense of loneliness. For many who have FM, that sense of aloneness is eased when we realize that there are nearly 10 million Americans who have fibromyalgia to some degree. You've got lots of company.

Anger generally follows. The worst aspect about being angry about FM is that there is no definite object to be angry with. If someone makes you mad, you can direct your anger at them. But in FM, there is no clear-cut culprit. No one knows yet why FM strikes, and although some can point to a particular incident that seemed to set it off, FM just creeps up on many others. There is no one and nothing to blame.

You might get angry at the doctors, but it's not their fault either. You may be angry because someone failed to diagnose FM sooner or refused to take you on as a patient, but doctors are only human too. Getting angry at them will only affect a relationship that has a tremendous impact on your well-being.

The third step is usually a period of bargaining. You begin to believe that if only you act in some specified way, then everything will be okay. You try to bargain your FM away. Most people realize fairly quickly that this approach is not going to work.

Depression usually follows bargaining. Depression can become very serious, and very insidious. You have the right to feel sad, and it is while coping with depression that you come face to face with the reality of your chronic condition. You are most vulnerable during the depression stage.

Sometimes the depression is minor, particularly if your FM is mild or in remission. At other times it can suck all of the light and happiness from your life. Depression is a stage at which you need to maintain a constant support system. Quite often it occurs for only a short time, but there are instances when it blocks out any potential for a happy future.

Signs of Depression

- Feelings of sadness that last far too long
- Major changes in sleep habits, resulting in more or less sleep
- Listlessness
- Poor concentration
- Major changes in eating habits
- Sense of worthlessness
- Severe feelings of guilt
- Lack of interest in sex
- Thoughts of or attempts at suicide

It is important for you to recognize the symptoms of depression, to realize that they are a natural part of the grieving process and of living with a chronic health problem. Realize now that you can build a support group that will help you when it seems as though tomorrow and a better day will never come.

I want to include one word here about suicide. A couple of years ago, there was quite a bit of publicity because a woman who had chronic fatigue syndrome committed suicide with the assistance of a certain well-known doctor. There was a great outcry throughout the FM community because too many people felt that individuals obtaining a diagnosis of FM or CFS would think there was no hope for them and no future and might be tempted to follow this path.

I want to speak up because I have been at that place in the past and but for God's interference and a loving friend's care, I would have died. I had been in unbearable pain for more than six weeks with no relief. The few pain medicines that I could take didn't faze it, I had no insurance, and on top of that, the Veterans Administration felt my pain wasn't sufficient reason to admit me to the hospital. One evening I took an overdose of my Elavil. I have no memory of the paramedics or the emergency room. When I woke up in intensive care, I told my doctor that I hadn't really tried to commit suicide— and I hadn't. I just wanted to find some relief from my pain and I didn't care if I died to get it.

But when I woke up, I knew that I had been given a second chance and I wasn't about to mess it up. I went from the local hospital straight to the VA hospital in Dallas. The doctor in the emergency unit told me my pain still wasn't reason enough to admit me. I told him that I had nearly died and I was not leaving that hospital until they did something to help me. So they sent me to the psychiatric ward. Trust me, you don't want to go to such a place. But I did learn some important things. The doctors worked with me on pain control and put me on medicine for both my depression and pain. I continue to see a psychiatrist to help control my ongoing depression with antidepressants in higher doses than are given for FM. But the factor that helped my depression the most was finally getting my disability from the VA—something that didn't happen for 7 more years—and relieving my financial worries.

Why am I telling this? Because I know that I am not alone. I have had a number of letters and phone calls from individuals who have read earlier editions of this book and realized that 1.) they weren't crazy; 2.) they weren't alone; and 3.) there was some relief for their pain. I have also learned another very important lesson from that incident. I will never continue to push myself until my pain gets so completely out of control that it is impossible to bear. I still have pain and there are times when I think, "I can't handle this." But I remember November 10, 1987, and I know that yes, I can handle it and I will do all that is within my power to ease my pain and I will *not* continue to do things that keep my pain building. If that means I go to bed and stay there all day or for a couple of days (Just wait until this revision is finished, I may stay there a month!), then that is what I will do. If it means that I have to buy my own muscle relaxer at $66 a bottle because the VA will only prescribe Soma now, and I can't pay some other bill, then so be it.

SUICIDE IS NOT AN OPTION! My pain has never been that bad since, and I believe it was that bad because I kept pushing, trying to keep a job and go to college when my body had simply reached the point when it could no longer handle that. Let me tell you, life has been sweet since then, even with the sourness of FM. Even though I still have pain, sometimes very bad pain, I still choose to live.

If you are depressed or if you feel you just can't handle your pain anymore, find a friend, a family member, your doctor, or another health care provider or a suicide hotline. Find someone to help you through the Valley of the Shadow of Death, because there is still life with FM. Despite the pain, the fatigue, the frustrations and problems, it can still be good. After all, if I had died on November 10, 1987, the first edition of this book would never have been written, much less the second and now third edition. You never know what may come out of your own experience of pain, maybe something even more wonderful. But you have to grab someone's hand when you hit that dark valley until you can come out on the other side.

When we grieve for a lost loved one, we eventually reach a stage of acceptance. We know that person is gone forever and that we must go on. In your grieving process, you will also arrive at a state of acceptance, although it will not be quite as permanent as it is when losing a loved one.

Quite often, your grief will be cyclical. There are times when you will feel you have come to terms with the reality of life with FM and won't expect to feel anger, guilt, or depression again. But then a time may come when you want desperately to do something and realize you can't, because the price you will pay in pain, fatigue, and in time lost at work is too high. And so you will return to the cycle of anger and depression.

But you learn to overcome the emotions that go with grief. No one can stay wrapped up in them forever. Once you have made your first trip through those emotions, you know that you can handle whatever comes—you know that you'll live through them and emerge from the experience a stronger person.

ELEVEN

How an Occupational Therapist Can Help

I first met Jean Judy when I was looking for some help in coping with my FM. Ms. Judy, who was then a registered occupational therapist and an assistant professor of occupational therapy at Texas Woman's University in Denton, Texas, helped me in many ways. Her students helped me make the sign that reads "I'd rather be riding a horse" for the back of my wheelchair. And it was she who introduced me to a craft that was not painful to do and that became a new way for me to make money. Do not overlook the possibility of occupational therapy when considering available resources for help.

An occupational therapist's goal is to assist people in continuing to fill the roles they have chosen in life. To accomplish that, a two-objective plan is developed for each role. For a person with fibromyalgia, the objectives are to conserve energy and to develop any adaptions necessary for continuing the chosen role. Due to the inconsistent nature of fibromyalgia, these plans are very individualized.

For people with mild fatigue and pain, for example, the occupational therapist's ideal goal might be to help them to continue working at a full-time job and maintaining their household responsibilities as well. The occupational therapist looks for ways to perform daily activities that require the least amount of energy and place the least amount of strain on the muscles over a 24-hour period, while also encouraging full range of motion in all major joints. Normally, the occupational therapist would conduct a thorough investigation of both the home and the work site. Based upon your individual needs, the occupational therapist suggests techniques and adaptations to the environment that would enable you to conserve energy, making it possible for you to fulfill your roles.

In an interview, Ms. Judy explained the role of an occupational therapist: "The occupational therapist strives for a balance of work, play, and rest. We believe that a person must have all three. Often, we are able to take an activity the individual was involved in previously and find a substitute that still gives that person satisfaction."

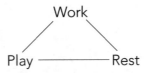

Generally, occupational therapy clinics are associated with hospitals or rehabilitation centers, although there are some that work with the elderly through home health agencies. Occupational therapy training centers often provide services for members of the community. Check with the colleges and training centers in your area, because they often offer services at discounted rates. The cost of occupational therapy services at these centers varies and is sometimes covered by insurance. Because occupational therapy clinics are not available everywhere and doctors don't always think about referring people with fibromyalgia to them, it may be hard to find one on your own. The information that follows is just a small part of what you can do to make life with FM easier. If you have access to such a clinic, make a point to check it out. You will be amazed by the amount of practical information that an occupational therapist can provide.

Many times in the past when a certain chore had become very tiring, I considered trying to find a way to simplify it. But because I've always felt that I wasn't really handicapped, I believed I had no excuse to consider making adaptive changes in my daily routine. Few people with FM think of themselves as handicapped and so view an occupational therapist's help as something they aren't entitled to. This approach is completely counterproductive. You don't have to think of yourself as handicapped. But accept that you do have limited energy and pain that often prevents you from doing what you want and from continuing to function. If occupational therapy can show you how to do necessary tasks while conserving enough energy so that you can still participate in pleasurable, if not essential, activities, then you must tap that resource. Occupational therapists have some extremely helpful ideas about how you can simplify your work, whether it is in the home or outside the home. Let's look at some of the areas in the house where changes can be made. These examples should teach you to look at your environment as if you were seeing it for the first time or from a stranger's viewpoint. Increased awareness will help you find ways of saving energy. There are many general tips that help make day-to-day living easier. Your doctors and therapists can suggest additional ideas to supplement these.

You Are Entitled

You don't have to be "handicapped" to take advantage of resources that will make tasks easier for you and that will give you energy to enjoy the more pleasant activities in life.

Planning

So where do you start? Begin with planning. If you take the time to plan out your day, your work at the office, and your housework, you will save yourself energy and pain in the long run. Take the time to sit down and write out the various roles that you play in your life. Are you a wife, father, banker, or teacher? Get a yellow legal pad and put one role at the top of each page.

Then on each page list the activities that you do for each role. Let's just start with being a parent. As a parent, you not only handle meal preparation and clothing needs, but you also coordinate your children's activities. Depending on the number and ages of your children, you can often get them to help in ways that will give you more energy. Youngsters can learn to pick up their toys and put them away, saving you from having to bend over 30 times a day to put toys in the toy box. And think about how heavy toddlers become. How many times a day do you bend over to pick up a small child without thinking about the best way to do it without causing pain? Learning to lift heavy objects correctly is essential in conserving energy and reducing pain.

If your children are older, they can help out around the house in any number of ways. Sit down together and discuss ways they can help you to conserve energy so you can keep the family functioning as normally as possible.

Priorities

It is important to set priorities. Sometimes that means not doing things you've always done. It may mean not pulling out the stove and refrigerator to scrub the floor except when you buy a new appliance or move. It may mean that instead of mopping and waxing the floor every week, you give it a lick and a promise during the week and once a month, you give it a top job.

As you are making out your list of activities, be sure to distinguish between tasks that must get done and tasks that you would like to get done. You will find that there are some jobs that you can skip over in order to conserve energy for more important activities.

Play

Even though people with FM need a balance of work, play, and rest, they often decide to ignore the importance of leisure activities. It happens most often when all of your energy is devoted to make a living. What happens to priorities then? You have to earn a living. But you are only hurting yourself in the long run if you don't include some play in your life. Many people may experience periods of remission, some will have days with less pain, but fibromyalgia is here to stay—don't burn yourself out today.

What kind of play will fit into your altered lifestyle? Occupational thera-pists try to come up with something that is linked to your activities before FM. For example, if you can no longer ride horses, can you still keep them around? Is there someone who can care for them? If not, have you considered photographing horses? These sorts of options are available for everyone. With your cooperation, your occupational therapist should be able to help you find outlets for pursuing your interests.

Pacing

To manage the fatigue that accompanies FM, part of your planning should involve pacing. It is generally easier to talk about pacing than to actually do it. When your mind is intensely focused on a project, you may not want to stop working on it unless you have to. Or it may be that you have always got-ten up early and jumped into the less pleasant chores so you could finish them and move on to more enjoyable activities.

For some it may be better to jump in and get it over with, while others will learn that pacing is a necessity. Pacing can be as simple as doing one household chore at a time, such as cleaning the cat box, and then taking a break.

Look at the overall picture. You may have to spread your work over sev-eral days in order to accomplish it at all. There will always be some chores that must be done at a particular time, but there may be others that can be shifted.

It is important to listen to your body. The goal, of course, is to do as much as you can and to live as full a life as possible. Sometimes that means work-ing on in spite of the pain and fatigue. At other times, it is necessary to rec-ognize when you need to slow down. Your body will generally give you plenty of warning when you need to slow down or take a break.

You may need to go to bed much earlier than you are accustomed. This may have an impact on your social life, but careful planning will allow you to maintain some sort of social interaction. Learn to recognize when you can push yourself and when you should rest.

Pacing should apply to the larger picture of your life as well. If you need to conserve your energy to generate income, that may mean that you don't

Rest and also walk. I try to walk a mile a day. I take Vistaril at night and that is working good. (I) get rest at night. I take the herb Cat's Claw (Una de Gato) and that helps my immune system. I started a support group one and a half years ago. I get much more out of that. I have accepted that I have fibromyalgia; I think the sooner you do accept it the better off you are. —Marion

The 4 Ps

Planning

Priorities

Play

Pacing

do a lot of housework or cooking. Consider spending one day preparing meals to put in the freezer, allowing the time spent cooking daily meals to be used for other purposes. Then you only have to pop them in the microwave. If you still have the energy to cook, it doesn't take much more effort to prepare extra food for use at a later time.

Whatever your situation, pacing can help you maintain a more stable and pleasurable life. Don't dismiss the potential benefits of pacing—you'll be surprised how much it helps to restore peace to your life.

Living with Children

If you have young children, try to eliminate bending and stooping as much as possible. This is especially important because most research indicates that FM quite often affects the lower back and hips. Try using plastic milk crates as carryalls. Even if your children are too young to understand the concept of putting their toys away, you can often make the activity into a game, having them put the toys into the crate, which can then be carried to their room or to the toy box.

Even with cooperative children, your work may be eased by using such adaptations as an extender. An extender has a long handle that helps you to reach and pick up not only items on the floor but also those above you. It may be just what you need to save a lot of bending and stooping.

A parent may pick up a baby or a toddler over and over again. Although it is often quickest to bend over and pick up the child, stop and think about what stress you are putting on your back. If possible find a chair or stool and sit down first, then pick up the child. If it isn't possible, use your knees and thighs instead of your back to squat down. You may need to steady yourself by holding on to a piece of furniture as you come back but that's much better than putting all of the load on your back. This holds true for picking up any heavy or awkward item.

In the Kitchen

Everyone has to eat. If you have teenagers in your family, you might set up a cooking schedule that operates on a rotating basis. Even younger teens can put together a pizza. This arrangement will benefit you and your family; it will generate more family time and prompt your children to learn valuable cooking skills. Cleanup time can be shortened if you have an automatic dishwasher, although some people with FM don't mind washing dishes because the warm water makes their hands feel better.

Although take-out food can become very expensive, even if you are single or live with only one other person, it is an option worth considering. The quality and nutritional value of take-out food is steadily improving. Allow yourself the luxury of turning to take-out food when you might have dismissed the option before.

> *I find that when chopping vegetables, using a large chef's knife is less of a strain, even though small knives are easier to wash.*
>
> —*Harriet*

If you like to cook or must cook for a family, there are still some things you can do to make it easier for yourself. Get an office chair with wheels for your kitchen. Whenever you are having trouble standing or getting up and down, use your chair to scoot around the kitchen. Try doing most of your preparations at the table. If you like to use the counter as your workstation, get a bar stool with a good back on it. Sitting down while doing many tasks is a proven energy saver. Use lightweight dishes and pans. Heavy objects add to the strain on your wrists and hands; you may even drop them. The same thing goes for pitchers. Heavy ceramic or ironstone pitchers may be attractive and match your dishes, but they're extremely hard to pick up and hold.

Store items you use most often in convenient places at levels that don't require you to do a lot of extra stooping and stretching. Try to reduce the weight of drawers by removing items you rarely use and storing them elsewhere. A lot of this may seem obvious, but many people arrange their kitchens the way their mothers did or the same way they've always done it. Rethink the way your kitchen is organized. What arrangements can you make that will save you steps and energy?

Handling Laundry

Laundry is another taxing chore that plagues everyone, but it can be especially difficult for those with a limited range of motion. However, there are a number of ways you can make laundry less difficult.

If you have to make a number of trips out to the laundry room, consider having two laundry baskets that stay in your bathroom or bedroom—one for clean clothes and the other for dirty ones. Whenever the dirty one is full, you can carry it to the laundry room. Folding laundry, particularly sheets and towels, requires holding up one's arms while performing repetitive motions that cause the muscles in the arms and shoulders to burn and ache. Consider leaving clean sheets, towels, and other items unfolded in the other basket so you can pull them out whenever you need them. If you have a large family, use several baskets, designated for whites, colors, towels, and other batches of laundry. Place them in the laundry room, if you have room, so you won't have to carry them back and forth. Make each member of the family responsible for placing his or her dirty clothes in the right basket. This will free you from sorting and transporting clothes.

The same idea can be used for items that need to be regularly picked up and returned to their proper place. Set an attractive basket or plastic milk crate in each room. When you finish with something, whether it's a book or some papers, drop it into the crate, and once a week or as often as necessary, make one trip to return the items to their particular place. (If you don't already have a system that provides a designated place for everything, this may not be the best time to establish one. Sometime when you are feeling especially good, ask a friend or family member to help you reorganize. Not only do more hands make the work easier, but it's a great opportunity to visit.)

Making the bed can be difficult. Make up one side of the bed at a time and, if possible, make sure the bed is clear on both sides so you don't have to do any more stretching and reaching than absolutely necessary. If you choose to fold your sheets, put them in sets, wrapping one in the other and tucking the pillow cases inside. This way you'll have to reach into the closet only one time for each bed.

Plastic closet organizers, which have no drawers to pull open, are great for keeping commonly used clothes handy and neat. The items you use most often should be easiest to get to.

Ironing demands a lot of energy from the arms and legs. You might want to avoid all cotton garments and stick to cotton blends. They combine the benefits of easy care and comfort. If you can afford it , take your clothes to the dry cleaner and allot the saved energy to a more enjoyable activity.

Managing Bathroom Chores

With a little creativity you can do several things to ease the burden of bathroom chores. Like most people with muscle pain, you may find that you have a hard time reaching around the bathtub to clean it. Try cutting down a mop handle until it is a length that allows you to easily reach all around the tub. Next, cut a large sponge in half, and then cut a hole in the middle of one of the halves. Fit the half with the hole on the end of the mop handle, making a versatile scrubbing tool that won't force you to get down on your hands and knees to clean the tub. It is also good idea to quickly run the washcloth over the fixtures and around the flat areas as you leave the bathtub. This is a great habit to develop because it takes only a few seconds once you get down into the tub and saves you from having to return later. Do the same thing at the sink after washing your face or brushing your teeth.

If you have cats and have trouble keeping the litter pans clean, you might consider this: Cut a large plastic trash bag into two pieces and lay one piece on the floor. Set the clean box on top of that, placing the second piece of plastic in the bottom of the box. You will need a cat box that has a rim to fit over the sides and hold the plastic liner in. (It's a lot cheaper to use a plastic trash bag than to buy special liners.)

Because sacks of litter are generally heavy, don't lift them and pour litter into the pan. Instead, open the top of the bag and use a large plastic margarine bowl to scoop out the fresh litter. You should also have a "pooper scooper." Try to set the box close to the toilet. Whenever you're in the bathroom, scoop out the solid waste, dump it into the toilet, and flush it away. If your septic system can't handle the solid waste with the used litter, put the waste in plastic bags you've saved from shopping trips. Tie the neck into a knot and drop in your trash that is going outside immediately. This way you will have to change the entire box only once a week. Keep the plastic bags, a broom, and scoop in the bathroom to save steps. Long-handled scoops further decrease the amount of bending and stooping you'll have to do to clean the litter box. If you can afford it you may consider buying one of the new "self-cleaning" litter boxes.

Juggling Shopping, Errands, and Transportation

You may have always enjoyed shopping and doing errands but now find that such activity leaves you almost sick with fatigue. Getting in and out of the car and walking around stores may be more than you can manage. The same is true for going to art galleries, museums, and special events. It's bad enough to have to give up something you enjoy, but it is more difficult to accept that you are unable to take care of essential business.

You may find yourself reaching a point where you postpone going to the grocery store or running any other errand because it just takes too much energy. If you live alone, it may be necessary to rely on friends to do your grocery shopping and to use the mail as much as possible for other business. You also might consider asking your doctor for a letter that will allow you to obtain a handicap license plate or sticker. The letter will state that your mobility is impaired and your condition is permanent. You may get some unpleas-

> When people ask you what fibromyalgia is, it's very hard to define it. The best way is to describe it as "very many painful spots in my body" and "worse than arthritis." Also, if you find a doctor or therapist who understands it, stick with him/her. —Mary

ant comments from people who don't see anything obviously wrong with you, but people with FM are not the only ones with "invisible disabilities." Don't take it personally; there are far too many people who do not have a health problem who park in those spots. At least *you* have a legitimate reason to. Being able to park close to the door may enable you to get around the grocery store or post office without generating debilitating pain.

Alternatives to Getting Around

What can you do if you still can't get around? It would be easy to give up and stay home, but you would become completely isolated from the world. Consider some of the alternatives. Ask to use the wheelchairs or three-wheeled electric scooters that a number of stores provide. Many department, discount, and grocery stores have wheelchairs with large baskets, and many also offer electric scooters to customers, available on a first-come, first-served basis. Check with your local merchants.

Consider renting a wheelchair for special events if you have someone to push it. You should consult your doctor before using a wheelchair for more than special events. You may have no other option than to use a wheelchair, but you should never make the decision on your own. Those who use a wheelchair regularly know that it is very important to stretch their hip and leg joints each day. Not doing so is dangerous because it allows the muscles to shorten, further limiting the range of motion. The object of using a wheelchair periodically is to remain active and not to become totally disabled, which is possible if you decide to use the wheelchair full-time.

A motorized three-wheel scooter requires some energy to use, but you will be able to do so much more than you could without it. Cheaper models run under $2,000 and can be taken apart and carried in the trunk of your car. Before making the investment, be sure you can take one apart yourself and lift each piece unless you have someone else who can do it for you. Lifts can help you get the scooter in and out of your car, but at $1,200 or more, the cost of a lift is significant. Not all scooters will fit in all cars, so shop carefully.

Even without investing in a three-wheeler, there are several ways you can keep up with your errands and shopping. Many grocery stores offer a delivery service for a nominal fee and it's not just for those who have trouble getting around. Many provide the service because it is time-saving for people who work. Check with the stores in your area to find those that provide delivery service and what their fees are. In most communities there are individuals who will run errands, either for pay or on volunteer basis, for those who can't get out. Look for their ads in your community's newspapers or contact your local United Way or Chamber of Commerce.

There are other simple ways of conserving energy. Let your fingers do the walking! Do all comparison shopping over the telephone, especially when pricing prescriptions. Try to pay all of your bills by mail. Save heavy shopping, whether for groceries or other goods, until you have help.

Catalogs are also a great way to shop. I love to read but find it hard to get to the bookstores as often as I would like. There are several catalogs that carry popular fiction books as well as those that focus on nonfiction or reference books. There are catalogs for almost anything you could want, including other catalogs.

One of the newest ways to shop is on-line with your computer and modem. Look into Prodigy, America-Online, CompuServe, and the world wide web through your Internet Service Provider.

When shopping, you will generally find that store employees are very helpful in getting items off the top shelves or dividing up purchases so that no one bag is too heavy. Try not to feel self-conscious or hesitant about asking for help; it will save you from extra discomfort in the end. Today most stores offer a choice between paper and plastic bags. Although plastic bags are environmentally the better choice (if recycled), they often put a greater strain on your hands and forearms than do paper bags.

Don't Be a Shut-In

Continuing to drive helps those with FM maintain some independence, but driving and riding often increases the stiffness. In fact, many people with FM become very uncomfortable when they travel for any distance. Any extended travel is usually so painful for them that they have to limit it as much as possible. This can keep them from seeing family members as often as in the past, or from taking part in special occasions.

When you must travel, make frequent stops and allow yourself enough time to walk out the stiffness. If you make sure you are as rested as possible when you start out, you may not have to completely give up travel by car. If you don't have access to a large car with a good suspension system and comfortable seats for such trips, consider renting one . You will find that you'll be less tired and enjoy your stay more if you take these steps to make your travel comfortable.

> *When feeling "tired" and "sore" and "ill," to help everyone around me from being blasted out of the water, I go to my bathroom, light candles, turn off the lights, and soak my tightness away! —Toni*

One thought while we're discussing getting out and about: If you aren't up to going out, encourage your spouse or other family member to go out anyway. Just as in a relationship where both partners are healthy, each person needs some time and space for himself or herself. When a couple must deal with the stresses of a chronic illness, this becomes even more important. And it is compounded if the partner with the chronic illness wants the other person to live and move at that slower pace dictated by the illness. This is not an easy path to walk. It requires compromise and careful thought so that the needs of both individuals are recognized and met.

Public Transportation and Travel

Public transportation can be extremely difficult for someone with FM. Although taxis can take you door to door, they are generally expensive. Buses and trains are more difficult to use. Many public transportation systems have special features to make them accessible to disabled people, but you may still have to walk some distance to and from them. Some organizations provide rides for those with handicaps. Check your local United Way or Chamber of Commerce for information.

Traveling by air can be an exhausting ordeal, but there are techniques you can use to help ease the trip. If you are traveling alone, make arrangements to use handicapped parking and inquire about using a wheelchair. Your airline will have a wheelchair waiting to meet you and take you to your gate. Generally, you will be allowed to board first, although you will possibly disembark last. Despite the risk of losing checked luggage, you should carry on as little of your luggage as possible. You are going to be tired at the end of your trip; consider having someone drop you off and pick you up or taking a door-to-door shuttle.

When traveling make all arrangements ahead of time, taking advantage of handicapped parking and any other special services the airline provides. If you drive to the airport and want to bring your three-wheel scooter, most airlines will check the scooter on through, transfer you to one of their wheelchairs, and then deliver the scooter to you when you reach your destination. The most important thing to remember is to plan ahead. Make your arrangements as early as you can so you can take advantage of any help that will save you energy.

In the fall of 1995, I joined a group touring Regency England. I took along my scooter but also made arrangements to have a regular wheelchair waiting at our first hotel. Because electrical service is different in England and Europe, I got a converter so I could charge the scooter. But something went wrong anyway. I had two days before it quit and the rest of the trip was spent using the wheelchair. By the time I returned to Texas, I was exhausted and in a good deal of pain but I have never regretted it for a moment.

Everyone on the trip, from other members of the tour (all strangers—at first) to the wonderful coach driver to hotel staff, were wonderfully helpful. They went out of their way to help me out, whether it was to push me or help me carry some of the "goodies" I collected along the way. I am a book lover and I had a great deal of fun shopping, even when I couldn't get into the stores. England is a lot less accessible than the United States but the people were wonderful. You haven't lived until you and your wheelchair have been pushed up two very narrow, very steep ramp strips into the Bank of England or until a delightful quartet of Frenchmen have lifted you and your chair down some stairs into the lobby of a London hotel. I was asked more than once if the trip was worth the pain and exhaustion and I always say "Yes, yes and always yes." Next time I go, I'm just going to make arrangements to rent an electric scooter over there. I'm told it's very expensive in London because they aren't stored within the city; the price of real estate for storage room costs too much. But I want to go back to the British Museum, the bookstores in Bath, and so much more. If you want to go badly enough or need to do something enough, you'll find a way.

Self Care

There are a number of things that you can do to make taking care of yourself somewhat easier. If you find that you can't hold your arms up long enough to continue your daily grooming routine, you need to think about adopting a "look" that requires less effort. Men may opt to grow a beard, eliminating the strain caused by shaving. Women might consider another hairstyle—one that doesn't require the use of a curling iron or rollers. Try a casual style that is easy to care for, yet flattering. Your self-confidence is already damaged—pick a look that you like and feel good about. Don't assume you'll have to sacrifice your attractiveness for ease.

In the Bathroom

There can be several other difficulties involved in taking care of yourself. Because of muscular pain, you may have a hard time getting up and down to use a toilet. When you're out in public places, always take advantage of the handicap stalls. Taking a bath often requires some advance planning to determine the best and easiest way of getting in and out of the tub. Without becoming overly anxious about using the tub, try to be more aware of the dangers of falling. Put no-skid strips on the bottom of the tub. Install a bar on the wall or side of the tub that you can hold on to when getting in and out of the tub. Get a short stool to sit on while in the tub, so that you won't have as much trouble getting out.

If you have a shower, you might consider using a shower stool and secure support bars. If you put the bars in yourself, make sure they are connected to the studs behind the wall and not just to the wall itself. The bar must hold your weight without coming out of the wall if you slip and start to fall.

If there are activities that you can no longer manage to do, such as shav-

I use my floor as my masso-therapist. I roll, vacillate and gently strike parts of my body that are in the pain. Create. Write poetry or jokes or create art that expresses the pain. Changing awareness of it to a different part of the brain actually eliminates much of the pain during the activity. —Lynn

ing the back of your legs or washing your hair when your neck and shoulders hurt so bad that the very thought of bending over makes you shudder, there are some steps you can take to help. Try adding a handle to your razor. It can be something as simple as part of a curtain rod fastened to your razor handle with strapping tape. You need to measure first to get the length you need, and you may need to keep a second razor handy for areas that you can still reach easily.

When washing your hair, try pushing a chair up to the sink so you can kneel and take some of the strain off your muscles. Even having your hair washed at a salon may lead to some uncomfortable moments. Dortha always puts an extra towel beneath my neck because it hurts to lay back while she washes my hair.

Consider bathing every other night instead of every day, unless you've been particularly active, and have as much as possible prepared ahead of time. Take everything you will need into the bathroom at some point earlier in the day.

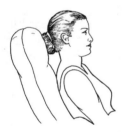

Make sure that your head has a good support. The head and neck are a prime area for pain. If you are going to be sitting in a chair for some time, try to get one that will let you rest your head against it, at least periodically.

A bar stool with a back on it will allow you to sit at the bathroom mirror to put on your makeup or shave. If you have a dress that zips up the back and has one spot where you just can't reach, get a piece of twine, tie a knot in one end, and run the other end through the hole in the zipper. Holding the twine, pull the zipper up, and slip the twine out of the hole. Repeat the process when you are ready to take off the dress.

Whenever possible, wear slip-on shoes, but not backless ones which don't provide enough support. Try to find athletic shoes which fasten with velcro instead of laces. Even if you aren't going running, they provide excellent support. If you must buy some with laces, try to get them with laces that are short enough to not trail on the ground. That's an invitation for a fall if you trip on them. If you have to wear neckties regularly, try loosening

The body pillows help make you more comfortable by easing the strain on the back and shoulder. Since those with FM have trouble getting restorative sleep, they need to do anything they can to improve the quality of the sleep they do get.

the knots, instead of completely untying them, so that you can pull your ties on or off over your head and tighten the knot, avoiding the added muscle strain involved in completely tying it.

Rest

Rest plays a vital part in the effort to conserve energy. But there are several important aspects of resting. Some sort of break should become a part of your regular schedule. At home, you might try a variety of different methods. At least one session of relaxation therapy or meditation would be a good start. Or you might just lean back in your recliner, if it takes some of the pressure off your lower back, and stay there for 10 to 15 minutes.

At other times, you might need to stretch out on the couch and actually fall asleep. When this happens, your body needs the extra rest, so don't try to fight the sleepiness. Take the phone off the hook or plug in the answering machine so you won't be disturbed. Taking regular short breaks, may reduce the need for long deep naps. Try not to take too many naps, because this may interfere with your nighttime sleep.

Comfortable and Supportive Furniture

Many people never really think about furniture and its comfort until they develop chronic pain. Sometimes it takes trial and error to find things that can make you more comfortable. Make sure

Again, when you are sitting, try to have the proper support for all parts of your body. Here, she has a pillow for her head, armrests, a hassock for her feet and a pillow in her lap to rest her book on, thus elevating it up a bit to improve her comfort.

Aerobic exercise and jazzercize, swimming (heated pool), humor, rest, good nutrition (organic), touch, massage, hugs, and affirmations. —Sharon

you have a good supportive mattress and a pillow that supports your neck rather than your head. Your doctor or physical therapist may be able to recommend a pillow that properly supports your neck. You may also want to purchase a foam rubber egg-carton or waffle pattern pad for your bed. If you are having trouble with your arms and shoulders, support them with extra pillows. When you lie on your side, the natural weight of the arm not supported by the mattress can pull it down, making you uncomfortable. Resting the arm on a spare pillow can significantly eases this discomfort. Body pillows are available to provide extra sleep support along the length of your body. These long, tubular pillows are available at most department and discount stores as well as healthcare outlets.

When sitting, elevate your feet. Put a footrest under the kitchen table, your office desk, and even your computer table. Anything from a cinder block to an upholstered stool works, but it should be about 5 or 6 inches high. If possible, sit in a recliner or use a footstool so you can get your feet up. When you sit on the toilet, prop your feet on an old set of bathroom scales or last year's telephone book .

It is extremely important that you have furniture that provides you with good support and lets you sit comfortably. A good bed promotes restorative sleep, and a good chair eases muscle strain by encouraging proper posture. By using comfortable furniture and frequently changing position, it is possible to avoid stiffness.

TWELVE

Establishing Support Systems

"No man is an island" is especially true for anyone who must face a chronic health condition. Those of us with chronic health problems may be more likely to fall into deep depression and believe we can't endure. But we don't have to face life with fibromyalgia alone. Sometimes we may feel that we would be better off alone, but that's not true at all. We need family and friends. And we need to know that others share our experiences.

The best support system includes family members, friends, and a support group, supplemented by such organizations as the Arthritis Foundation or one of the large fibromyalgia associations like Fibromyalgia Alliance of America or Fibromyalgia Association of Greater Washington. All of these groups play important roles in helping you maintain your well-being. Make an effort to establish such a system for yourself.

The Role of Family

Ideally, family is your first line of support. After all, these are the people who care most about you. It would be wonderful if family members always responded to our needs in a positive way. But quite often that doesn't happen. Spouses may feel angry and frustrated; children may not understand or resent the changes that have taken place.

Most of the time, the love and respect that develops during a marriage will carry a couple through the rough times. Unfortunately, the very nature of fibromyalgia puts a strain on even good marriages. Because FM is so unpredictable, many times plans have to be canceled or changed. When a couple starts out, they usually bring certain expectations to the marriage. If one

Exercise daily. Keep busy. Try to work part-time at least. Attend support group meetings. Laugh—watch lots of TV comedy shows.—Beth

member can no longer carry out those expectations, such as working and bringing in a full-time income, the goals of the marriage must be re-evaluated and often reset. Sometimes, even spouses can begin to doubt the severity of the pain and fatigue. Spouses may also resent having to assume responsibility for activities that their partners previously handled.

An interesting study, conducted a number of years ago, found that in a large percentage of cases, when women contracted rheumatoid arthritis after marriage, their husbands were unable to cope with the situation and left the marriage. This usually did not occur when the man developed rheumatoid arthritis. If the woman already had rheumatoid arthritis when the couple married, the problem did not arise as often.

> *Educate yourself, family, and friends about FM by reading materials, listening to tapes and watching videos. Make education of FM a continuing process. Learn and practice pacing and coping techniques. Join a support group.—Carol*

Perhaps because it is still so new, no such study has been conducted on those with fibromyalgia. But any chronic health condition adds stress to a marriage. If you think your marriage may be in trouble, seriously consider marital counseling. If the love and respect are strong enough, you, your spouse, and your psychologist will find a way to keep the marriage together. You need that sense of support and love even more now that you face FM.

Avoid developing a "sick role." A "sick role" is the term for using one's condition to manipulate others or to avoid fulfilling responsibilities. Sometimes this behavior is unintentional and is often reinforced by family members and friends who care about the person and want to protect their loved one from additional pain and suffering. The person with FM, as well as family and friends, must realize that while there are limitations, life must still be faced and lived responsibly. How can you avoid falling into a sick role? Increased awareness helps. If you are aware of the potential for it, you are less inclined to develop one.

Conscious or unconscious manipulation only complicates an already complex situation. A certain amount of pain and fatigue inevitably accompanies FM. But if you try to totally retreat from this pain, you will also retreat from the joys of life. You must find the fine line between living with and yet also, sometimes, ignoring the real limitations of your condition. This is not easy. It requires commitment from both partners as well as some "tough love." Quite often it takes the help of a skilled counselor and a lot of compromise and communication among all those involved.

If a marriage is already on shaky ground, a chronic condition may completely destabilize it. The pressure of dealing with chronic illness could be the

> *Just show up. Many times I have found that when I'm feeling espe-
> cially bad and don't want to go to an engagement, volunteer work,
> or support group meeting, I remember these words that I heard on
> an AA tape, and I push myself to "just show up." I end up feeling
> better physically and emotionally.—Cindy*

impetus that ends the marriage, or it could also become the catalyst for
rebuilding a marriage full of the love and caring it was founded on. Although
no one knows in advance which marriages might fail under the burden, there
are some steps you can take to keep a relationship together.

Communication and counseling, as already mentioned, provide a good
beginning. Just as education about fibromyalgia is important for the person
who is learning to live with FM, it is also important that the spouse and family members learn about it. Once a diagnosis is made and a treatment plan
developed, the family needs to become involved. Some activities and
responsibilities may need to be shifted, temporarily or permanently. Goals
and priorities should be reviewed. It is not necessary to immediately change
everything around you, but as time passes and you understand FM's long-
term impact better, you can discuss your options as a family.

Your children need to be involved in this process, but be careful not to
force them to assume an adult's role while they are still children. Many times
when a parent becomes ill, the oldest daughter is expected to assume her
mother's responsibilities and the oldest son becomes responsible for his
father's duties. If there is any way to avoid this pattern, you must find it. It is
one thing to have children perform some role in maintaining the home, and
it is another thing to pull or push them prematurely into adulthood. Talk it
over with your family, so all members are familiar with the desirable changes
and can work together to prevent damaging patterns from developing.

When establishing your support system, don't overlook the importance
of your family's role. Many people find that they never would have been able
to come to terms with their FM without the support of their loved ones.
Adversity often draws a family closer together, bringing out new facets of
each individual's personality that may not have been evident before. Turn to
your family for help; you may be surprised by the inspiration and strength
they give you.

The Importance of Friends

Thank God for friends. Friends are an important part of everyone's life, but
they are especially important if you are single. As with family members,
friends must work to maintain their relationships. More important, friends
may be more likely than family members to walk away from someone they

feel is using them. And yet, there are times when a friend will be there for you when a family member is not.

Communication is vital to people. We all want and need to discuss the things, good and bad, that are happening in our lives. If you have a family, it is probably natural to share with them. But if you live alone, you must reach out and communicate with someone. This is when the telephone can become a lifeline.

Even if you are stuck in bed or at home for a while, the telephone is your doorway to the world. It enables you to contact someone when you need some specific task done or when you just need someone to talk to.

You don't have to tell all of your friends all about your fibromyalgia. Some will want to know all about it, and others will want to know only a little. Be selective in how much you say. For some, it may be difficult to accept that you look good even though you feel terrible. When those you aren't particularly close to ask how you feel, chances are pretty good that they don't want to know the specific answer. Just say something like, "I'm feeling better today" or "I didn't have such a good night" and move on to another subject.

> *Exercise, walk, try to be drug free. Cut out salt and refined sugar. Eat lots of fish and good food. Watch caffeine and alcohol intake or totally ban them from your diet. Life is so very short. Enjoy it. It hurts but it won't hurt you to be active and happy. Get moving and start living.—Diane*

With those who really care and really want to know, you still must be careful to avoid boring them to tears or continually dumping your troubles on them. Try to follow a few rules about your friends. Do talk to them about your fibromyalgia, but try not to let it monopolize the conversation—there are many other things to talk about. They need to know that you care about them and want to know what is happening in their lives and careers and how their families are doing. If your friends are married, recognize that their spouses and families are their first priorities and try to always respect that.

If you need to talk to someone, try not to call on the same person too often. Your friends may insist that they don't mind, but you don't want them to feel that you call only to complain. If you need to talk, call one friend today and another one tomorrow or the next day. And always make an effort to call or meet your friends for casual conversation, as you would do in any healthy friendship.

Try not to impose on them for favors. If you can manage to take care of something on your own, do so. If you can't manage by yourself, ask for help only if it is convenient for another person to help you. You want to remain as independent as possible, but don't carry it too far.

My tip is to find a "fibro-buddy," so you will always have an empathetic ear, a partner-in-crime to help you get back on track with pacing, resting, and exercising. And someone to do a lot of laughing with—whether you need it or not!! —Judith

Many times your friends are glad to help, they just may not be sure what they can do. You will find that the world won't come to an end if a friend has to do something for you. Actually, it gives them an opportunity to express their love and caring for you. Don't deny them that expression. Try to remember the good graces of friendship. Always ask for help, don't command or demand. Say thank you and mean it. And whenever possible, do something for your friends in return. There will usually be something that you can do, even if it is only listening when they talk about their problems.

As you come to terms with your fibromyalgia and its effect on your life, you may feel that your friendships are one-sided, with you doing all the taking and none of the giving. But you should know that you have something to share as well. Just as you often need someone to listen, your friends and family members occasionally need a listener, too. For many people, talking about their problems is one of the most effective ways of coping with their experiences. That's not always true for others, but sometimes the need to talk to a sensitive listener does arise.

Even with fibromyalgia you should try to maintain as wide a range of interests as possible, and try to have friends who share some of those interests. By having more to talk about than just your health, you invite stimulating and enjoyable conversation.

Remember the old adage, "If you want to have a friend, you must be a friend." Even when it seems your world has narrowed to just yourself and your fibromyalgia, remember that those around you have something that is equally important in their lives.

Support Groups

No matter how much your family and friends love and support you, unless they have fibromyalgia, they really don't know what you are experiencing. They may have been with you through the frustrating months or years while you sought a diagnosis and treatment, but they can't experience the pain, fatigue, and emotional impact of chronic pain.

Be part of a support group. You make friends! Friends! Friends! Fibro-friends! They understand your bad days and are happy for your good days.—Jill

Joining a support group can connect you with people who have the same special needs that are not met by other support systems. The knowledge that you are not alone makes it much more bearable. When others have already walked the path that lies before you, they may be able to share information that will make your journey easier. This is true whether you are coping with a divorce, a career change, or a health condition like FM.

Some support group members need to talk with others and to vent their frustration. Frustration is a word commonly used, for good reason, among people with fibromyalgia. The problems of getting a correct diagnosis, an effective treatment, and facing the day-to-day challenges of living with FM generate a lot of frustration. For many people, talking eases those feelings, and just knowing others have the same experiences helps. On the other hand, some support group members may just want to get the group's newsletter and stay informed of the latest research information.

Many support groups across the country share and distribute a lot of information. They stay abreast of the increasing numbers of articles that appear in medical journals and general periodicals. Several major newsletters on fibromyalgia also provide information for support group members. Three excellent ones are published quarterly, *Fibromyalgia Network*, *Fibromyalgia Frontiers*, and *Fibromyalgia Times*. These newsletters and their organizations can be good sources directing you toward more information.

Unfortunately, support groups may create the potential for members to focus on negative aspects rather than on ways of improving their lives. When you meet to talk about your fibromyalgia (or any condition), there is a chance that the meeting will turn into a gripe session instead of a healthy exchange of positive feelings. There may be people who attend only to have the chance to talk about the severity of their case and to monopolize conversation.

Many doctors have been hesitant in the past to sponsor or recommend support groups because past experiences have proved unproductive. They don't want their patients to continually focus on the negative or to dwell on their problems, because it is unhealthy to do so. It is best, when first organizing a support group, to adopt some guidelines that will prevent each meeting from turning into a "pity party." Sometimes it isn't easy, especially when you have members who feel the need to control sessions, but if everyone is aware of the reasons for the guidelines, you can work together to prevent negative results from overwhelming positive.

Ron Burks, a psychologist in Wichita Falls, Texas, has this important piece of advice for anyone, whether they are in a support group or among their friends: "Get off the pity pot." It's okay to talk about your problems with someone. Sharing them can divide them in half. But put a limit to the amount of time you are going to feel sorry for yourself. For that time, allow yourself to be unhappy, to recognize your sorrow. Then move on to something more

Sample Guidelines For Fibromyalgia Support Group

Statement of Purpose

This group is designed to provide people with fibromyalgia and other interested people the opportunity to talk freely about problems, concerns, and frustrations, and to share information, encouragement, helpful hints, and support. Through your participation in this group, you can learn more about fibromyalgia, and get ideas about what can be done to ease its effect on your life.

What Happens At The Meetings?

Meetings are open to the public and may include talks, with question-and-answer sessions, given by various medical experts, general discussions among members, and special entertainment. Educational sessions may include films or seminars on problems of daily living. Generally, the meetings last about 1½ hours.

Guidelines For the Support Group

1. We are a group of people with a common bond, sharing our concerns, feelings, experiences, strength, and wisdom.
2. Our discussions are designed to foster positive attitudes and are directed toward solutions. We share our problems, but we do not dwell on them.
3. We listen, explore options, and express our feelings. We do not prescribe, diagnose, judge, or give medical advice.
4. We know that what we share is confidential.
5. Our leaders are not "the experts," and most sharing of ideas will come from us.
6. We each have the opportunity for equal talking time or the right to remain silent; we can share as much or as little as we want.
7. We actively listen when someone is talking and avoid interrupting and participating in side conversations.
8. We encourage "I" statements, so that everyone speaks in the first person. We stick to our own experiences and avoid generalities.
9. Our meetings supplement and do not replace medical care.
10. We do not promote quackery or provide specific medical advice.
11. We each share the responsibility for making the group run smoothly.
12. Having benefitted from the help of others, we recognize the need to offer our help to others.

Courtesy of The Fibromyalgia Association of Texas, Inc.

positive. After you've done this, life will be a whole lot sweeter and your friends will still want to come around. But most important of all, you will feel better emotionally. And you will want to be around yourself for the 24 hours a day, 365 days a year that you must live.

Nonprofit Organizations

Organizations such as the Arthritis Foundation in the United States and the Arthritis Society in Canada; the Fibromyalgia Association of Greater Washington, D.C.; the Fibromyalgia Alliance of America, Columbus, Ohio; the National Fibromyalgia Research Association, Salem, Oregon; the American Chronic Pain Association; and the California Network of Self-Help Centers are other sources you can approach for support.

Look through the resource listing in the Appendix. You will find a number of organizations that are working actively to increase awareness of fibromyalgia as well as raise funds for research on it. Most of the organizations also make information on fibromyalgia available to those who request it. Others, such as the National Fibromyalgia Research Association, have research funding and advocacy as their primary objectives.

Even though fibromyalgia is still relatively new, the Arthritis Foundation and the Arthritis Society include information on it in their regular publications. In the United States the national publication is the bi-monthly *Arthritis Today*. Area chapters generally publish newsletters that are also available to anyone interested. The Arthritis Foundation also prints small pamphlets that cover subjects of interest to those with FM.

The mission of the Arthritis Foundation is to support research to find the cure for and prevention of arthritis and to improve the quality of life for those affected by arthritis. It carries out this mission in a number of ways. The foundation sponsors the Arthritis Aquatic Program, self-help courses, and arthritis clinics for low-income people; provides a wide variety of literature, a physicians referral list, seminars, a speakers bureau, and support groups; and sometimes has a loan closet for home medical equipment. They also help with the "Kids on the Block Arthritis Program," which teaches school-age children about arthritis. People with FM can also use these services and programs.

The Arthritis Foundation also sponsors fund-raising activities to help cover their expenses and go toward research.

THIRTEEN

When Fibromyalgia Affects Your Work

Most people today work outside the home because they must meet financial needs and because such work provides them with a sense of satisfaction and self-worth. This is also true for anyone with fibromyalgia, but for many of these people, working is not so easy. When you wake up exhausted, there isn't a lot of energy for charging into the working world, even if the desire is there.

Regardless of the type of work, you must have enough energy to handle both the negative and positive aspects of it. Whether you are an assembly-line worker, a secretary, or an executive, you must manage demands on both your physical and your emotional health. There is no indication that fibromyalgia worsens with time, but sometimes there are aspects of the working world that make it seem as if it does. Sometimes fibromyalgia has very little impact on a person's ability to work, but sometimes it forces a person out of a current job or out of work completely.

A number of studies worldwide on fibromyalgia and work disability show that those who indicated they were unable to work ranged from 5 percent to 46 percent, but the figures of those actually drawing disability income were generally around 25 percent. While more individuals are awarded Social Security disability benefits, they have generally had to appeal to an administrative-level judge to receive them. There is a resistance to recognizing both chronic fatigue and fibromyalgia as disabling conditions because of the potentially enormous burden that this would put on the system. In fact, in Norway the government passed laws in 1991 and 1995 limiting disability benefits to conditions with scientific validity. That slowed down the numbers receiving disability for FM and CFS, but these numbers have begun to climb again since 1995.

In an ideal world, everyone would have adequate health care, but many Americans do not have health insurance or the money to pay for regular health care. Often, a person's condition worsens for lack of that care.

Although the health care crisis is drawing more and more attention, it is unrealistic to expect any quick solutions.

Whatever your situation, it is important to realize the importance of working, not only for the financial aspect but also for the feeling of self-worth that it provides. If your FM is so bad that you can no longer work at someone else's schedule, there are many ways you can still earn income—don't give up. If you are still employed, do as much as possible to continue working. There are three basic ways to handle your work situation: retain your present job, change jobs or careers, or find an alternate means of income if you must quit full-time work.

Retaining Your Job

Even if you enjoy your present job and your fibromyalgia allows you to keep your position, your employer may have misgivings about your capabilities. Although they may be extreme cases, some people have been forced out of their jobs. This is a reality that you must prepare for and possibly confront. Quite often, employers do not want to retain employees who have a high absentee rate or who are unable to produce at top speed.

Everyone would like to believe that employers recognize the value of retaining employees who have been with the business for some time and who know their jobs well. And everyone would like to believe that a company would be willing to make some effort to keep those employees by adjusting the working environment or the job structure. Such companies and employers do exist. Only you know if you work for one.

Telling your supervisor about your fibromyalgia could mean that you will be eased, or even forced, out of your job. But there is also a chance that you won't. If you have worked for a company for a long time, the company may be willing to work with you. It is a good idea to discreetly check on your company's past actions in similar cases if you can. If you have worked for the same company for some time, you may already know about your company's record in situations like yours. Be cautious about voicing your concerns too quickly. If you leave your job, do so whenever it is best for you. Let it be your decision, not the decision of your supervisor or the company's management.

Americans With Disabilities Act

If you are unsure about your legal rights and the way your company is handling your situation, read the Americans with Disabilities Act (ADA). It may help you retain or obtain employment. It addresses the issues of employment and public accommodations for those with disabilities. Employers with 25 or more employees were covered starting July 26, 1992, and those with 15 or more were covered beginning July 26, 1994. A booklet titled "The Americans

Preparation of the Disability Report For The Patient With Fibromyalgia

Because fibromyalgia is predominantly a pain syndrome in which there are no objective signs of severity, the physician needs to interpret the data for the reader. Readers expect objective evidence of impairment and need to be educated concerning the nature of the fibromyalgia pain syndrome. The following guidelines are suggested.

1. Establish that the diagnosis is present. Cite the historical and physical findings that support the diagnosis. Use the 1990 American College of Rheumatology criteria for the classification of fibromyalgia (15). Often many other specialists are involved. Confusion may arise because the patient may have been diagnosed by others with disc syndromes, osteoarthritis, rheumatoid arthritis, carpal tunnel syndrome, thoracic outlet syndrome, LE, etc. Explain how these disorders can be confused with fibromyalgia, and insist on the diagnosis of fibromyalgia. Diagnostic criteria are helpful since, in our experience, most other alternative diagnoses are made on the basis of assertion.

2. If there is an alleged injury, describe the pattern of the development of fibromyalgia after the injury and whether it seems representative. It is often useful to note whether the fibromyalgia seemed to arise de novo in an otherwise healthy individual.

3. Describe the severity of the syndrome. The tender point count, pain diagram, pain and severity scores, and psychological and functional disability data are essential. There are normative data for these measures in fibromyalgia and other rheumatic conditions. Citing these data in relationship to your patient helps to put into perspective your patient's plight and status. If you have longitudinal data, they may be very helpful in assessment, and they should be given.

4. Is the patient disabled? I usually answer the questions this way; I indicate that only the patient feels his pain, and that it would be difficult to do certain tasks if there were such pain. I state that most patients with fibromyalgia are able to work provided the job doesn't place undoable physical demands on them. I cite the Cathey et al. study which showed that 30 percent of patients changed jobs because of fibromyalgia, but that most patients were able to work (32). I indicate that the course of fibromyalgia is most often chronic.

Courtesy of Dr. Frederick Wolfe and The Hawthorne Press, Binghampton, NY and *Journal of Musculoskeletal Pain*, Vol. 1, #2, 1993, pages 83-84.

with Disabilities Act, Questions and Answers," published by the Civil Rights Division of the U.S. Department of Justice, explains the ADA's purpose in the following terms:

"The ADA prohibits discrimination in all employment practices, including job application procedures, hiring, firing, advancement, compensation, training, and other terms, conditions, and privileges of employment. It applies to recruitment, advertising, tenure, layoff, leave, fringe benefits, and all other employment-related activities.

"Employment discrimination is prohibited against 'qualified individuals with disabilities.' Persons discriminated against because they have a known association or relationship with a disabled individual also are protected. The ADA defines an 'individual with a disability' as a person who has a physical or mental impairment that substantially limits one or more major life activities, a record of such an impairment, or is regarded as having such an impairment.

"The first part of the definition makes clear that the ADA applies to persons who have substantial, as distinct from minor, impairments, and that these must be impairments that limit major life activities such as seeing, hearing, speaking, walking, breathing, performing manual tasks, learning, caring for oneself, and working.

"An individual with epilepsy, paralysis, a substantial or visual impairment, mental retardation, or a learning disability would be covered, but an individual with a minor, nonchronic condition of short duration, such as a sprain, infection, or broken limb, generally would not be covered.

"A qualified individual with a disability is a person who meets legitimate skill, experience, education, or other requirements of an employment position that he or she holds or seeks, and who can perform the 'essential functions' of the position with or without reasonable accommodation. Requiring the ability to perform 'essential' functions assures that an individual will not be considered unqualified simply because of inability to perform marginal or incidental job functions. If the employee is qualified to perform essential job functions except for limitations caused by a disability, the employer must consider whether the individual could perform these functions with a reasonable accommodation.

"Reasonable accommodation is any modification or adjustment to a job or the work environment that will enable a qualified applicant or employee with a disability to perform essential job functions. Reasonable accommodation also includes adjustments to assure that a qualified individual with a disability has the same rights and privileges in employment as nondisabled employees."

Making Your Job a Little Easier

Even without the ADA, there are many things you can do to make your present job easier. Again, this is something that is very individualized and depends on your job and your degree of FM, but here are some general ideas you can think about.

If you work in an office and are able to consult with an occupational therapist, ask him or her to visit your job site and evaluate it. Your occupational therapist will review the work environment first, identifying anything that could be easily changed. Do you work under an air-conditioning vent? This is something that could aggravate your FM. Can you or your desk be moved? Can a screen or partition be set up to block the direct flow without making everyone else in the office uncomfortable?

Is your chair comfortable and does it provide adequate support? What about your desk? Is it the proper height? Do you find yourself in a strained position as you work? Do you have a footrest under your desk to ease the strain on your leg and back muscles? Is there a chair you can sit on or a stool you can rest one foot on if you have to make more than a couple of photocopies? (Switch feet if you stand with one foot on a stool and one on the floor for even 5 minutes. This alternately relieves the strain on your legs and back muscles.)

Do you have a lot of walking to do? Can you plan your day's activities so you won't have to move around a lot or sit in one position for too long? Unfortunately, every job cannot be planned in advance. Many times activity in the workplace is based on uncontrollable factors such as customer's needs, telephone calls, and those little emergencies that seem to pop up continually.

Is there any way that present equipment can be adapted to ease the strain on muscles? If you spend a lot of time on the telephone, think about getting a phone rest, a speaker phone, or a headset.

Two types of strain—physical and emotional or stress-related—can occur in the work environment. Although physical strain might be eased by using a more comfortable chair or moving the furniture, emotional, or stress-related strain, is not as easy to avoid. Generally, this type of strain is generated by the demands of the job or by your coworkers. Because stress is one of the factors

Sitting at a computer for long periods of time can increase your pain level and repetitive motions also can cause pain. Make sure you have a chair that provides good support, including support for your arms. Use a wrist supporter for the computer; some come with the computer, others can be purchased separately. You might even consider one of the ergonomic keyboards, but try them out before you buy. Make sure the monitor is at eye level to prevent neck strain. Also use a foot rest of some sort, as this helps alleviate some of the strain on your back.

that can exacerbate the symptoms of FM, you may need to re-evaluate your job and your commitment to it.

If you are constantly challenged by your job, find it exciting, and don't feel exhausted every night, don't leave it. A boring job, even if it is easier, would most likely cause more problems. But if you find that you can't maintain the daily pace, that you are completely drained every night, or that you snarl at the kids each evening, you need to think about changing jobs.

Sometimes it is possible to move to a position that generates less stress without leaving your employer. Depending on your career goals, you may feel left out of the corporate challenge. This may be a case where you have to reevaluate your priorities.

Quite often, stress at work comes from coworkers. Many times office politics and interpersonal relationships can make you miserable. It seems like there is always someone who continually passes the buck or who likes to keep the office stirred up about something. If you face such a situation, you must determine how much this type of stress affects you and your FM. Keep in mind, however, that this sort of stress will often exist where two or more people work, so changing jobs will not always eliminate it.

Making the Transition from Full-time to Part-time

It may become necessary to give up full-time work and take a part-time job. Adjusting their hours and the demands of the job will enable many people to continue working and earning money, even if it is on a lesser scale. That might be the best thing for you, particularly if your FM has become seriously debilitating. The change to part-time work allows you to keep your foot in the employment door and set your own pace, so that FM does not have such a major impact on you and your lifestyle.

If you can afford it and have a family, this may be one way to increase the time and energy you have for your family. By easing some of the physical strain of full-time work, you can schedule rest periods and activities with your family.

Often, it is also possible to leave a stressful full-time job for a part-time job that is much less demanding. There are many part-time positions that provide income yet do not have the stress that a full-time job does. Whether you decide to continue to work part-time depends on your needs. If you are not facing serious financial problems, part-time work may give you the best of both worlds by allowing you to earn money without having to cope with the added pain and fatigue your full-time work caused. However, be aware that part-time workers are not usually offered the same health-care benefits that full-time workers are. Make sure you understand the terms before you make any major change in your employment.

Changing Jobs

What if you have tried, but have been unable to keep your job, because the company is unwilling or unable to work with you or because your fibromyalgia is just too severe to let you continue in that job?

One possibility is to take some time off work to take care of yourself and possibly to find a new job. Another option is retraining. This is probably going to be necessary if you have been doing physical labor that you can no longer do. How and where do you find retraining? What does it cost?

Depending on your financial circumstances, you might consider going to a community college or university and making a complete career change. Even though recent budget cuts have affected many of the programs that offer grants and loans to students going to college, there is still a lot of help out there, and it's possible that you may qualify for such help. Most colleges and universities also have a center for handicapped students. These provide assistance needed to meet class requirements and to access all facilities.

You must realize that pursuing a new educational goal often adds to the stress level. If you haven't been to school in some time, the period of adjustment may be especially stressful. For some time now, most colleges and universities have seen an increase in older students attending classes for the first time or returning to the classroom so that they can change careers. Most institutions are aware of the difficulties older students encounter and will generally work to ease the transition for those students. But only you can decide if you are able to handle the requirements of attending classes at a university or a community college. If you decide to return to school, take advantage of the institution's career counseling. Most career counselors are great information sources and will be glad to help you find new employment opportunities.

If your fibromyalgia is serious enough, use your state's vocational rehabilitation agency as a resource. The goal of these agencies is to help the handicapped secure and maintain jobs. Those eligible for this help have disabilities ranging from orthopedic deformities to mental health problems, alcoholism or drug addiction, internal medical conditions, and more.

Not everyone with FM will be eligible, but if you have had problems working because of your fibromyalgia, you should contact the nearest office to determine your eligibility. The rehabilitation centers' services may include a medical, psychological, and vocational evaluation to determine the nature and level of your disability, job skills, and capabilities; counseling and guidance to help you and your family plan proper vocational goals; limited medical treatment if it is possible to mend the disability; some assisting devices that will stabilize or improve your function on the job and at home; training in a trade school, business school, college, university, rehabilitation center, or even on the job; and selective job placement services. Check with your state rehabilitation program for eligibility requirements and services.

Fibromyalgia is a condition that may present some problems for a person seeking help. The lack of documentation, such as lab tests and x-rays, is one of the problems. Because of this, it is very important that you have complete reports from your doctor documenting your illness and any job loss that has occurred due to FM. It may be that with the research now being done on the HPA axis, a test will be found that can be done without expensive equipment or processing. Another problem is that FM often has periods of increased pain when work may not be possible, followed by relatively pain-free periods. Each case is decided on an individual basis. If you believe that you qualify, or even if you think you do, contact your local office or state rehabilitation program.

Turning to Self-employment

Even if you can't work at a part-time job, there are ways to earn money. For the person who is willing to try new things and to learn, there is always a means of earning some money. You must be realistic and understand that it is unlikely you'll be able to make enough to replace your full-time income. Most importantly, you must be willing and you must maintain a positive attitude so that you can recognize opportunities and take advantage of them.

One of the best aspects of self-employment is that it allows you to set your own schedule so you can pace yourself. Sometimes you must push yourself, but then you will be able to take the time afterward to recover. Remember, fibromyalgia does not cause your body to deteriorate; you will not be damaging joints and muscles by being active. You will have limited endurance and may pay for your actions with some increased pain. But compare that with not earning any income, and you will discover that even though it hurts, you will have gained more than you've lost.

How do you find work you can do that will earn you some money? First you need to sit down and look at yourself and the skills you already have. Not only do you have to consider your education, work experience, and interests, you must also evaluate the strength of the economy in your region. Realistically assess the market for your services or goods.

Take a step-by-step approach to this process. You may want to enlist a career counselor to help you do this. First, list your education. What do you already have that can be useful in self-employment? Is there any way you can use your skills to work from home, at your own pace?

If you have at least a master's degree, you may be able to teach as an adjunct instructor at a local college or university. Pay, of course, varies with the college and the locale. Teaching as an adjunct is not quite self-employment, but usually if you are not teaching more than one class and your students have access to you before or after class, the only time you are officially committed to is the actual class time. All of your preparation and grading work can be done at your own pace.

If you do not have the education required, sometimes work experience is considered adequate. Many junior and private colleges hire people from within the community to provide instruction based on their hands-on experience in a specific field.

If there are no openings with colleges in your area or you don't have enough education, there are a number of other ways that you can still teach. They do not usually pay as well, but they do provide some income. Check into continuing education departments at colleges and public schools, as well as municipal recreation departments. These offer classes in almost anything that they believe people are interested in from writing and photography to calligraphy, from flying an airplane to gourmet cooking.

If you have sufficient experience, and can establish a reputation, you may be able to conduct classes on your own. Even in small towns there are people who wish to learn new skills or arts. There are classes in writing, music, theater or drama, dog obedience, personal finances, painting, games, dancing, and quilting. If you have a special skill, it's likely that others in your community are willing to pay to learn it. Of course, you must set reasonable prices and a clear agenda for the classes. You may have to rent a hotel meeting room to hold your classes, but you may also be able to use your church or civic organization's facilities for a minimal fee.

Although many people believe that writing books is for someone else, you may have special knowledge or experience that you can share with others. The ratio of nonfiction to fiction books published each year is about 40,000 to 5,000, or eight to one. Many of those nonfiction books share personal experiences; others share knowledge. Take the time to go to a well-stocked bookstore and do research. People are hungry for knowledge. What special knowledge do you have to share?

Even if you do not have the ability to teach others or to write, there are many ways that you can earn money. Many hobbyists have earned a living by expanding their activities and creating goods to sell. Many craft malls display and sell handmade items for a flat rental fee without requiring you to spend long hours at a booth. If you can tolerate spending a day at a booth or table, many local and regional fairs charge an entry fee and provide space where craftspeople can sell their work. You must obtain a tax number from the state comptroller's office because you must collect and pay sales tax on the items you sell. But by doing so you receive a certificate that enables you to buy your supplies without paying sales tax and perhaps even buy them at a wholesale or reduced rate. Many people earn enough money during the pre-Christmas period to live on for an entire year.

Some other ways of generating income include these possibilities: establish a secretarial, typing, or bookkeeping business at home; start a video or photography business; produce a newsletter for a church, organization, or club; begin breeding purebred dogs or cats; set up a genealogy service;

serve as a consultant or do freelance work, sharing your knowledge directly. Working in direct sales is another way of earning money. For some people, selling cosmetics, soft goods, or jewelry is an enjoyable, low-stress method of drawing an income. The most important thing to remember about self-employment is that there will be no time clock to punch and no supervisor peering over your shoulder. To earn money, you must be motivated and a self-starter. It is possible to fit this in as a way of working in spite of your fibromyalgia but you must learn to know when to rest and when to push yourself.

Check with your local Chamber of Commerce for information about SCORE, the Service Corps of Retired Executives. They offer free assistance to those who are considering going into business for themselves. Another source for help is the Small Business Administration.

What If You Need Help?

Sometimes it turns out that no matter how hard you try, the threat of financial problems becomes overwhelming. Where can you turn for help? As discouraging as your situation may seem, there are some resources that you might turn to.

Check to see if you have a disability insurance policy through your place of employment. Generally, these policies cover only a limited period, but that may be all you need to get back on your feet. Carefully read any such policy; don't immediately dismiss it just because it appears to offer limited coverage. If you have the opportunity to buy disability insurance through your present job, read it very carefully to ensure that you will be paid. Insurance disability policies can be very vague or limited, to your cost if you need them.

Apply for Social Security disability benefits. As previously explained, it is difficult for those with fibromyalgia to qualify for benefits. You may be turned down on your initial application but appeal it; most cases for FM have been won at the administrative law judge level but only if you have a judge who is knowledgeable about FM. If not, bury them in documentation and experts' testimony. Sometimes even then the judge may rule against your expert but don't give up. Each case is judged individually, and they do take into account each applicant's level of education and line of work. Even if you don't think you have much of a chance of getting the benefits, apply anyway and take it as far as you can.

What if you are like many people with fibromyalgia who slip through the cracks? You are able to earn some money but not enough to live on, and that amount disqualifies you for many forms of assistance. If you find yourself at this point, you should know that there are still some resources available and people who want to help.

If you're worried about paying your basic bills, there are a couple of sources that can help you locate funds to pay the electric bill before they cut it off or the rent before you are evicted. Although finding those agencies and associations that are willing to help can be a frustrating search, it doesn't have to be.

On the local level, approach the United Way office, information and referral services, or the Chamber of Commerce office to acquire a list of human service agencies. Check the telephone book for other human service organizations in your area. Many phone books list such agencies in a special section in the front of the book.

If you belong to a church or temple talk to your pastor, rabbi or religious leader. Although some churches still provide help on an individual basis, many have turned their contributions over to an interdenominational ministries program. Contact the social welfare office at your local hospital; you don't have to be a patient to get referrals from them for sources of assistance.

Depending on the region, there may be additional programs available to you. You can be directed to them by contacting the regional department of the human services office. Each state has a department of human services with regional offices throughout the state. The personnel at these offices can put you in touch with programs originating with federal, state, county, or city agencies. Each program has its own set of guidelines and eligibility requirements.

Asking for and accepting help, whether it is transportation to the grocery store or to the doctor's office or money to pay this month's rent is often very difficult. But many of these programs were set up to help those who are unable to manage, particularly because of their health. There is nothing that says you will have to use their help forever. But at least during the rough times, try to find out if you qualify for assistance. Later on when you are back on your feet, the funds can be passed on to someone else in need.

FOURTEEN

What You Need to Know

What is fibromyalgia?

Fibromyalgia is a chronic medical condition with widespread pain which generally affects the muscles, ligaments, and tendons. It is not a rheumatic condition, although rheumatologists are the medical specialists who most often diagnosis it. The word syndrome, which applies to the group of signs and symptoms, is also used because at this point we don't know exactly what causes fibromyalgia. It does not fit into the usual disease pattern so it cannot be considered an illness.

What are some of the other names fibromyalgia has been called?

Fibrositis, fibromyositis, tension myalgia, myofibrositis, interstitial myofibrositis, myofacial pain syndrome, myofascitis, musculoskeletal rheumatis, muscular rheumatism, nonarticular rheumatism, tension rheumatism, psychogenic rheumatism, psychophysiologic musculoskeletal reaction, and conversion reaction.

What are the symptoms?

The primary symptom is widespread pain that often feels as if it comes from the joints, but generally comes from the muscles and the locations where ligaments attach muscles to the bones. Pain, like the aching that accompanies a bad case of the flu, can occur all over the body in addition to the presence of sharp pain in specific spots. The words that have been used to describe the pain are quite extensive including: burning, sharp, dull, aching, radiating, deep. Tender points at specific locations around the body are extremely sensitive. Individuals can indeed feel pain "all over," in hands, arms, legs, feet, fingers, shoulders, hips, legs, knees and chest, and even in some internal organs. Other symptoms include fatigue, stiffness, non-restorative sleep, hypermobility, dry, itchy or blotchy skin, ocular complaints, such as difficulty in focusing, cognitive dysfunction problems such as difficulty in finding a particular word, poor concentration and short term memory loss, hypersensitivities to noises, light, and odors, and disequilibrium.

What are tender points? Are they the same as trigger points?
Tender points are the places where the ligaments attach the muscles to bone. Trigger points are spots that will cause pain at another location when pressed. (This is called *referred pain*.) Although the terms were used interchangeably at one time, they are not the same.

What are some of the conditions that can accompany FM?
While it was thought for some time that individuals with FM "just happened" to also have some of the following conditions, more recent theories posit that the symptoms of these conditions may also be caused by whatever perpetuates FM. Not everyone develops all of the following conditions. Some people experience some of them, some of the time. They include chronic fatigue syndrome, migraine or tension headaches, irritable bowel syndrome, irritable bladder/interstitial cystitis, painful menstruation, depression, anxiety, heartburn/gastroesophageal reflux, Gulf War syndrome, Temporomandibular Joint Dysfunction, Mitral Valve Prolapse, multiple chemical sensitivities, regional fibromyalgia/myofascial pain syndrome, and dryness of the mucous membranes in the nose, mouth, and other locations.

Can a person have fibromyalgia and a form of arthritis?
Yes. Some people with fibromyalgia also have rheumatoid arthritis, osteoarthritis, lupus, or other types of arthritis.

What affects FM?
Generally, the symptoms are made worse by damp, cold weather, sitting in a draft or in front of an air conditioner, both positive and negative stress, overexertion, or inactivity. Warm, dry weather, warm baths, mild exercise and rest may ease the symptoms.

Who gets FM?
Studies are still being conducted on the demographics of FM, but at this time we do know that more women than men get FM (about 85–90%), and generally it affects people in their 30s to their 80s. However, it has been found in children as young as five. It is also possible that many adults first experienced FM when they were young, but were unaware of it at the time or may have received the diagnosis of "growing pains."

What causes FM?
At this time, the definite cause is not known. However, the onset can sometimes be traced to an accident, an emotional trauma, or a viral-like illness. But in other cases, the onset is gradual, making it difficult to know when symptoms first appeared. Currently there are several theories as to the cause, most with a link to the central nervous system. These theories include a pain perception problem, neurochemical factors, allodynia, muscle abnormalities which can often be traced back to the central nervous system, inconstant serotonin levels, a neurotransmitter which carries messages through the body, and elevated levels of substance P in the cerebrospinal fluid.

Why do I feel so tired?
People with FM often have regular disruptions in their sleep patterns. Studies

have shown that the deepest level (stage four) of sleep is often disrupted. It is during this stage of sleep that the body is restored. But for those with FM, this deep sleep is continually disrupted as if there were an alarm constantly going off, preventing the sleeper from awakening refreshed. The symptoms of FM—pain, stiffness, and fatigue—have been reproduced in healthy subjects by continually disturbing their stage-four sleep.

Is fibromyalgia contagious?

No.

Does FM run in the family? Can my children inherit it?

At this point, that is unclear, although there have been many cases in which more than one member of a family has FM, and some studies have shown a possible tendency for it to run in families. At this time, no specific genetic link has been found.

What tests are used to diagnose FM?

Previously nothing showed up on either routine lab work or x-rays. Now because of extensive research, there are some tests which can be performed. However, the diagnosis is still made by the physical symptoms and history, and the elmination of other disorders with the same symptoms. Generally, there is no need for numerous, expensive tests.

Is FM crippling or life-threatening?

No, but the pain and fatigue can become disabling. Everyone's case is different. Some people experience only minor fatigue and pain, which can be treated with mild analgesics. Others find it difficult to carry on their normal daily activities. Recent figures indicate that about 30% of those with FM are on some form of disability or consider themselves unable to work because of their symptoms. Almost 50% of those surveyed have made some sort of adjustment in their working conditions or job, but continue to work. There are a large number of women who are considered housewives and, therefore, don't fall under the definition of disabled.

What can I expect in the long run?

Studies have not found FM to be progressive, but it is adversely affected by such factors as stress and extreme physical activity or inactivity. Like many other chronic conditions the symptoms vary from day to day and month to month, but have been determined to remain fairly stable over the long run. People who have had a few years of treatment still report pain and fatigue. Remissions can and do occur, but, again, that varies with each person. A remission is considered to be at least a two-month period without pain and stiffness. Some remission periods have lasted years.

Is FM the same disease as chronic fatigue syndrome/chronic fatigue immune dysfunction syndrome (CFIDS)?

Some researchers believe that there is a group of patients who meet the criteria for both fibromyalgia and chronic fatigue syndrome. There are also some, such as Dr. Muhammad Yunus and Dr. Daniel Clauw, who believe that there are a number of conditions which are caused by whatever causes the FM.

What kind of doctor should I see for treatment?
Generally, a rheumatologist treats fibromyalgia, but as more and more family physicians and internal medicine physicians learn about FM, they can also treat you. It is not uncommon for individuals to see orthopaedics, physiatrists (those who treat pain) and now even, chiropractic physicians, homeopathic physicians, acupuncturists, and other alternative medicine practitioners.

How can I find a doctor who treats fibromyalgia?
Contact your local chapter of the Arthritis Foundation or Arthritis Society. These organizations normally maintain a list of doctors. If there is no chapter near you, contact your medical society or hospital for a list of doctors. There are also some medical referral lists available from some of the fibromyalgia associations.

What kind of treatment can I expect?
Every person should be treated individually. At this time, tricyclic antidepressants such as Endep and Elavil are commonly prescribed to improve sleep. These are given in much lower dosages than when they are prescribed for depression. Also muscle relaxers like Flexeril, which is also a tricyclic medication, may be prescribed. Sometimes a nonsteroid, anti-inflammatory drug may be prescribed for pain if regular aspirin or acetaminophen (Tylenol) does not work. Some doctors will also prescribe mild narcotic analgesics if your pain is severe.

What other kinds of treatment can I expect?
Often a doctor will refer you to a physical therapist, who may use heat or cold to relax the muscles. A physical therapist may also use ultrasound treatments and may teach you some stretching and aerobic exercises that you can do at home. Cognitive behavioral therapy is also an important part of the treatment plan. This can range from relaxation therapy to stress management and is discussed under the section on psychologists.

What can I expect from my doctor?
The first step is to develop a good doctor-patient relationship. This means that you must be willing to take an active part in your case. Your doctor should confirm your diagnosis, providing assurance that the condition is not life threatening or crippling. Treatment for FM can be frustrating for both doctor and patient, and often you will want your doctor to spend some time with you. That is your right as a patient, but remember that your doctor has other patients as well. Do not expect a magic cure. FM is a chronic condition.

What other health care providers should I see?
Your doctor may refer you to a physical therapist, as already mentioned, but you may also want to see an occupational therapist, a social worker, and a psychologist or psychiatrist.

How do I find these health care professionals?
Any of these specialists' services may be available at your local hospital. Even if you have not been hospitalized, you may still use some of these services. Physical therapists, psychologists, and psychiatrists in private practice can

also see you. Generally, your doctor will refer you. Even if you have not been hospitalized, you may contact the social worker on staff at a hospital for referrals and assistance with financial and other needs. Occupational therapists may also be affiliated with universities or medical schools that offer an occupational therapy degree program, although many are now opening their own private practices.

What does an occupational therapist do?
The function of an occupational therapist is to find ways to assist you in continuing your activities in your chosen roles. The therapist can evaluate your home and job environment to find ways to conserve energy and ease muscle strain. He or she can also help find or develop adaptive devices that will help you continue to function.

What does a social worker do?
A social worker helps people find ways to handle financial and social problems.

What does a psychologist do for a person with FM?
Psychologists provide counseling services to people with FM who are grieving for the loss of their health, who are angry about the changes in their lifestyles, or who are experiencing other emotional problems brought on by a chronic illness. They can also assist family members in coping with the changes FM has caused. Sometimes they also offer stress management, biofeedback, self-hypnosis, or relaxation therapy training.

My doctor believes my symptoms are all in my head. What can I do?
Unfortunately, there are still some doctors who do not recognize FM as a legitimate physical condition. Instead, they consider it a psychological problem and try to treat it accordingly. Continue to look until you find a doctor who believes in fibromyalgia and who is willing to work with you on your care.

How can I learn more about FM?
As more and more research is conducted, more information is becoming available. If there is a medical school library near you, check the listings in *Index Medicus*, which covers most of the medical journals published today. Another source of information is a support group or association that provides such information. Check with the Arthritis Foundation or the Arthritis Society. The Arthritis Foundation publishes a bi-monthly magazine, which is available to anyone who makes a donation of $20 or more. Addresses for the Arthritis Foundation and the Arthritis Society chapter offices and for a number of fibromyalgia associations and support groups are listed in the appendix. You may also subscribe to "Fibromyalgia Network", a quarterly newsletter. There are several associations which produce excellent material on FM (see the appendix). Another place is the Internet. Just be aware that there is a tremendous amount of information out there on the Net, and not all of it is accurate.

What is a support group?
A support group is usually a collection of people who share the same inter-

est in a topic, such as fibromyalgia. They work to provide information and emotional support for each other when others outside the group are unable to. They may or may not meet on a regular basis, publish a newsletter or maintain a library of material on FM.

How do I find a support group?
The appendix lists a number of support groups around the country, but there may be one in your area that isn't included, as information changes often. Your local Arthritis Foundation chapter can provide helpful information, and a national directory of services which includes support groups has recently been published by the Fibromyalgia Association of Greater Washington (FMAGW). Contact your doctor, hospital, local medical society, or newspaper to find one. If there is not a group in your area, you might consider organizing one yourself. Contact the California Network of Self-Help Centers, the Canadian Council on Social Development, the Arthritis Foundation, or the Arthritis Society (see the appendix for their addresses) for assistance in starting a self-help group.

Can I donate blood?
The answer is usually yes, but it is best to check with your own physician first.

What research is being done on fibromyalgia?
At this time, more and more physicians and research centers are becoming interested in fibromyalgia. As the search for the cause of FM continues, there seem to be two directions of research, a central mechanism and a local or muscular mechanism. Researchers are looking into the central nervous system (CNS) with focus on the neurohormone, endocrine, and immune systems.

Another group believes that the answer lies within the highly sensitized pain perception of the central nervous system. This is becoming a major area of research as more and more abnormalities are found within the central nervous system. Often these changes within the CNS are believed to have an impact on the muscles, tendons and ligaments.

How can I cope with my pain if I can't use pain pills?
The use of narcotic analgesics, or pain pills, in the treatment of chronic pain—whether for FM or terminal cancer—has become quite controversial. Many people, including physicians, believe that the individuals given narcotics are in danger of developing an addiction to the drugs. The entire subject of chronic pain is being reevaluated as far as treatment is concerned. More doctors are willing to work with a patient in controlling the pain by medicine, but there are also many who recommend alternative methods of pain therapy including distraction, relaxation therapy, self-hypnosis, and biofeedback.

What if I can't maintain my level of work or if I can't keep working at my job?
There are several options that you might consider. The first is to attempt to

retain your job. You might consider some changes that will make your job easier. However, not all supervisors and employers are willing to make such changes, even though the Americans With Disabilities Act was passed to help individuals, including those with FM. The law hasn't been in place long enough to judge how effective the law will be.

You may need to cut down from full-time work to a part-time job that will put less strain and stress on you. Stress can make FM's symptoms worse. You might become self-employed, so that you can pace yourself according to your FM.

If you have to make a drastic change in your employment situation, you may be eligible for vocational rehabilitation. The appendix lists addresses for each state's vocational rehabilitation services office. Contact your state office to find branch nearest you.

If I can't work, can I draw Social Security Disability?

The Social Security Administration is still having a difficult time handling applicants with fibromyalgia. There have been some claims that the Social Security Administration won't add fibromyalgia to its list of disabilities for fear of overwhelming the system. Whether true or not, the number of those applicants who have been granted benefits is very small. Often, fibromyalgia is not listed as the cause. Instead, it may be listed as psychological or arthritic, even if these are incorrect diagnoses. When it comes down to it, I would rather receive benefits under an incorrect diagnosis—even if the system refused to recognize the correct diagnosis and its debilitating effect—than not receive them at all. The determination of disability is affected by the applicant's age and education, and if they are not expected to be able to work for at least twelve months.

What can I do to help increase research on fibromyalgia?

Because the competition for research dollars is so severe and fibromyalgia is still a new area, the amount spent on research for FM is much lower than for other health conditions. It is important that you let your Congressional representatives know how you feel about fibromyalgia and its impact on your life. Write, fax, or email them to urge them to increase money for FM research.

What do I tell my family and friends about FM?

You might have your family members read this book. For younger members of the family, explain that you have to make some changes in your lifestyle because of FM. They need to know that it is not life threatening, but there will be days when you are in pain and may not be able to do some of the things you had planned.

As for how much you explain to friends, that depends on how close you are. Some might also be interested in reading this book. But for every friend who wants to learn about FM, there will be an acquaintance who doesn't realize that FM is a chronic condition that fluctuates from day to day.

FIFTEEN

A Final Word

And now I come to the end of this third edition and it is with renewed hope that I look forward to the years to come. When I wrote the first two editions, I was depressed when I finished them because so little was known about fibromyalgia. As I have said several times in this edition, we still have a way to go before we have a cure for FM. But we are so much closer than we were in 1991 when I first researched the book. We have diagnostic criteria established by the American College of Rheumatology. The National Institutes of Health has set specific funds aside for research on FM, and although the amounts are not what we would like, they are a start. As we complete the work on this book, the NIH has issued a call for research study proposals. In his address to the House of Representatives Committee for funding research, Dr. Stephen Katz, Director of the National Institute of Arthritis and Musculoskeletal and Skin Diseases (NIAMS), specifically talked about fibromyalgia.

We've had articles on fibromyalgia appear in large numbers over the last few years in many of the medical journals, as well as numerous articles in women's and general magazines. Although, I still get the comments, "What is fibromyalgia? I've never heard of it," from both medical personnel and individuals I encounter in my everyday life, more and more often, I also get the response, "Oh, my (mother, sister, aunt, cousin, husband, neighbor, friend) has that, or was just diagnosed with it." The word is spreading, and more people are getting the correct diagnosis. Hopefully, more of them are getting that diagnosis without having to endure years of mis-diagnosis and or disbelief in their conditions.

Treatment is still focused on relieving symptoms and is generally only about 30% effective. But with many individuals turning to alternative or complementary medicine, there is some increased improvement. It's not a cure or a 100% successful treatment of all of the symptoms, but it is progress. It is perhaps, in the combination of traditional and alternative treatments, that we see many who do obtain relief from their pain.

For myself, the recognition of the mind/body/spirit link is very important. I know that there isn't a completely effective physical treatment, although the combination of the antidepressant, muscle-relaxer, pain relievers (both over-the-counter and prescription such as Tylenol with codeine #3), and anti-anxiety medications provide me with a lot more control over the symptoms while massage therapy brings me significant relief when my pain gets bad.

But it is in the mind that I feel I have more control. I do use relaxation therapy, meditation, and often creative visualization to bring my tensed-up and exhausted body to a point where I can allow the medications to do their work. And, perhaps the most important of all, I know that even in times when there is not enough relief from these methods, I can always choose how I will react to my pain, fatigue, and the physical limitations they place upon me. I may get angry, but I don't allow it to gain control of me, and the same goes for my frustration and depression. I have accepted that these are part of my life but I do not have to allow them to take total control of it. My mind is my ultimate weapon and I have the spiritual faith in God that enables me to fight my symptoms when necessary and to just accept them as a part of my life.

I would urge you to share with me the hope of increased knowledge that the researchers are striving to bring us, the hope that they will find the cause or causes and an effective treatment that will give us even more control over the symptoms. I would also urge you to do everything within your power to achieve the best life you can have under the present conditions, knowledge about FM, and the treatment. Above all, I would urge you to find some way in which you can use this chronic illness to grow personally, spiritually, and mentally.

At this point in time, you cannot change the fact that you have FM and you have limited means of treating the symptoms. But you, and only you, can control how you will respond to its reality within your life. It is up to you to educate yourself about FM and any of the associated conditions that you may have. With your doctor as your partner, develop the best combination of treatments and methods you find here or elsewhere to continue to live as positive, productive, and meaningful a life as you can. That is what you can control.

I wish you the very best.

Appendix

HEALTH CARE PROFESSIONALS

Acupuncturists are practitioners of a branch of Chinese medicine in which needles are inserted into a patient's skin as therapy for various disorders or to induce anesthesia.

Chiropractors (D.C.) practice a system of diagnosis and treatment based on the belief that many diseases are caused by pressure on nerves due to misalignments (subluxations) of the spinal column and that such diseases can be treated by correction (e.g. by massage) of the misalignment.

Counselors and psychologists provide psychological and emotional support and guidance while helping people and their families cope with chronic illness, depression, and stress. They may be affiliated with a medical facility, have a private practice, or work with a local human services agency.

Dietitians/nutritionists provide assistance on special dietary or nutritional needs and food preparation.

Family physicians have advanced training in family medicine and provide medical care for adults and children. When needed, they can refer patients to specialists.

Homeopathy is medical system based on the idea that "like cures like," which uses drugs or other substances that would produce in healthy persons the symptoms shown by the sick person (e.g., treating a fever by giving small doses of a drug that raises body temperature).

Internists have advanced training in internal medicine and in caring for diseases in adults. They can also refer their patients to specialists.

Medical doctors (M.D.) are physicians who practice *allopathy*, a system of medicine that aims to produce (by means of drugs, etc.) a condition opposite to or antagonistic to that affecting the ill person. These are traditional Western doctors.

Massage therapists are licensed by the state to provide massage therapy, which is defined as a scientific technique causing relaxation, stimulation, easing of mental and physical tension, alleviation of aches and pains, the breaking up of fatty tissues and muscle spasms, and the improvement of circulation throughout the body.

Nurses work in many environments from the hospital to doctors' offices and private homes. They can assist your doctor in providing you with health care. There are now many **Certified Nurse Practitioners** who have graduate-level degrees and can provide a higher level of health care than can either Registered Nurses (RNs) or Licensed Vocational Nurses (LVNs).

Occupational therapists are registered practitioners who also require a referral from your doctor. They work to help you to conserve energy and ease strain on muscles as you fulfill the day-to-day activities that your various roles require. They can evaluate your work and home environments as well as help guide you in your leisure activities.

Osteopathic doctors (D.O.) are doctors with the same amount of training as M.D.s and who use all of the usual forms of medical therapy, including drugs and surgery, but place greater emphasis on the relationship of organs and the musculoskeletal system and use manipulation to correct structural problems.

Pharmacists fill your doctor's prescriptions and can provide information about a drug's side effects and the way it responds when mixed with other medications. They can also advise you about over-the-counter medications.

Physiatrists are doctors who specialize in rehabilitation and treat symptoms with the help of such agents as heat, cold, water, electricity, and mechanical apparatus.

Physical therapists are licensed practitioners who work with you on your physician's orders. They can teach you exercises that will help maintain muscle strength and provide relief without using medication for pain control. They also work with heat and cold to ease pain and improve movement of joints and muscles.

Psychiatrists are physicians who treat and prevent mental disorders. Unlike **Psychologists**, psychiatrists can prescribe medication for such problems.

Rheumatologists specialize in the treatment of arthritis and rheumatic diseases.

Social workers can be affiliated with hospitals or medical centers, where they work to help people handle financial and social problems. They may provide counseling or assistance in learning coping skills.

Vocational rehabilitation counselors have special training to help those with physical and mental difficulties retain their jobs, obtain work, or reenter the work force.

Note: Not everyone will require the services of all the health care providers listed.

ARTHRITIS FOUNDATION CHAPTERS

These fibromyalgia support groups are sponsored by the Arthritis Foundation Chapters. Because there is a time lag from the time I compiled this information and the date of publication, there may be additions or deletions to this list. If in doubt, or if you are unable to reach someone at the address or phone number provided, contact the Arthritis Foundation Office listed before the support group, the nearest Chapter Office, or the National Office. (Area code changes may also affect the phone numbers.)

The Arthritis Foundation offers a number of excellent publications on a variety of subjects that impact the lives of those with fibromyalgia as well as other types of arthritis. I have not listed them all in the publications section because of space limitations, so contact the nearest Chapter, Branch, or Office for a list of their publications, check their web site or call the toll-free number for the National Office. They also provide speakers, special exercise programs, aquatic exercise programs, and other services. Check into the seven-week Fibromyalgia Self Help Course, taught by trained volunteers.

National Office
1330 West Peachtree
Atlanta, GA 30309
(404) 872–7100
(800) 283–7800
www.arthritis.org

Arthritis Today magazine
Arthritis 101 $11.95
250 Tips For Making Life With Arthritis Easier $9.95
Your Personal Guide to Living Well with Fibromyalgia $14.95

Alabama
Alabama Chapter
300 Vestavia Parkway, Suite 3500
Birmingham, AL 35216
Phone: (205) 979–5700
Toll–free: (800) 879–7896
Fax: (205) 979–4172
alchaf@aol.com

Dothan Office
211 North Alice Street
Dothan, AL 36303
Phone: (334) 793–2671
Fax: (334) 793–9057

Huntsville Office
316 Longwood Drive
Huntsville, AL 35801
Phone: (205) 533–7722
Fax: (205) 533–9074

Mobile Office
P.O. Box 161278
#1 Infirmary Circle
Mobile, AL 36607
Phone: (334) 432–7171
Fax: (334) 433–9125

Montgomery Office
310 North Hull Street
Montgomery, AL 36104
Phone: (334) 832–4003
Fax: (334) 832–4008

Arizona
Central Arizona Chapter
777 East Missouri Avenue,
 Suite 119
Phoenix, AZ 85014
Phone: (602) 264–7679
Toll-free: (800) 477–7679
Fax: (602) 264–0563
cazaf@unidial.com

Southern Arizona Chapter
6464 East Grant Road
Tucson, AZ 85715
Phone: (520) 290–9090
Toll-free: (800) 444–5426
Fax: (602) 290–0652
artfdn@azstarnet.com

Arkansas
Arkansas Chapter
6213 Lee Avenue
Little Rock, AR 72205
Phone: (501) 664–7242
Toll-free: (800) 482–8858
Fax: (501) 664–6588
arkansasaf@aol.com

FM Self Help Courses available
Support Groups:
Batesville, AP & L Reddy Room,
 793–6352
Bella Vista: Community Church,
 855–2435
Benton: Saline County Hospital,
 778–6054
Bentonville: Bentonville Police
 Department Conference Room,
 (501) 359–2104
Cabot: Mt. Carmel Baptist Church,
 843–6327
Camden: Ouachita City Medical
 Center, Conference Room C,
 836–8101
Cave City: Bank of Cave City, 283–5728
Clarksville: First Presbyterian Church,
 212 College Avenue, 229–3189
Clinton: Van Buren County Library,
 745–5791

Conway: Faulkner County Library,
450–7726

Crossett: Family Chiropractic Clinic,
(870) 364–8257

El Dorado: YWCA, 862–5442

Fayetteville: Washington Regional
Medical Center, Health Education
Center, (501) 443–0692

Fort Smith: Fort Smith Wellness Center,
(501) 452–0771

Fort Smith: Northside Community
Health Center, 783–2247

Gentry: Total Life Center, Seventh Day
Adventist Church, 736–8927

Green Forest: Community Room,
438–6283

Hardy: Arkansas Physical Therapy
Center, (870) 856–4126

Heber Springs: First Electric Building,
362–8530

Hope: Medical Park Hospital, 887–2145

Jonesboro: St. Bernard's Medical
Center, 972–5224

Little Rock: Pulaski Heights Methodist
Church, 771–1456 (evenings)

Marshall: Methodist Church, 447–2321

Mountain Home: Redeemer Lutheran
Church, 425–7085

North Little Rock: St. Vincent North
Rehab Hospital, 1st Floor
In-Service Classroom, (501)
834–1800

Pine Bluff: National Medical Rentals,
Community Room, 879–3060

Rogers: Rogers Public Library,
636–0897

Russellville: private home, 229–3189

Searcy: Central Arkansas Hospital,
268–1287

Stuttgart: SE Arkansas Behavioral
Health Clinic, 673–3507

Texarkana: Rural Electric Coop
Building, East 9th Street,
(501) 779–6031

Northeast Arkansas Office
Northeast Arkansas Rehabilitation
Hospital
1201 Fleming Avenue
Jonesboro, AR 72401
Phone: (870) 933–5142

California
Northeastern California/Northern
Nevada
3040 Explorer Drive, Suite 1
Sacramento, CA 95827
Phone: (916) 368–5599
Toll-free: (800) 571–3456
Fax: (916) 368–5596

Northern California Chapter
203 Willow Street, Suite 201
San Francisco, CA 94019
Phone: (415) 673–6882
Toll-free: (800) 464–6240
Fax: (415) 673–4101

Southern California Chapter
4311 Wilshire Boulevard, Suite 530
Los Angeles, CA 90010
Phone: (213) 954–5750
Toll-free: (800) 954–2873 (CA & NV
only)
Fax: (213) 954–5790
Has active FM support groups

Coachella Valley Branch
73–710 Fred Waring Drive, Suite 104
Palm Desert, CA 92260
Phone: (760) 773–3076
Fax: (760) 773–0858

Inland Empire Branch
11728 Magnolia Avenue, Suite B
Riverside, CA 92503–4958
Phone: (909) 688–6700
Fax: (909) 688–9729

Orange County/Long Beach Branch
17155 Newhope Street, Suite A
Fountain Valley, CA 92708
Phone: (714) 436–1623
Toll-free: (800) 452–2832
Fax: (714) 436–1625

Santa Barbara Branch
2942 De La Vina
Santa Barbara, CA 93105
Phone: (805) 687–1592
Fax: (805) 687–1594

San Luis Obispo & Kern County Branch
3220 S. Higuera Street, Suite 307
San Luis Obispo, CA 93401
Phone: (805) 541–1721
Phone: (805) 395–1450 (Kern County)
Toll-free: (800) 549–3153
Fax: (805) 781–3165

Valley Branch
16633 Ventura Boulevard, Suite 550
Encino, CA 91436
Phone: (818) 995–7378
Fax: (818) 995–6416

Ventura County Office
400 Mobile, Suite B7
Camarillo, CA 93010
Phone: (805) 389–5222
Fax: (805) 389–5225

San Diego Chapter
9089 Clairemont Mesa Boulevard
Suite 300
San Diego, CA 92123
Phone: (619) 492–1090
Toll-free: (800) 422–8885
Fax: (619) 492–9248

Colorado
Rocky Mountain Chapter
2280 South Albion Street
Denver, CO 80222
Phone: (303) 756–8622
Toll-free: (800) 475–6447
Fax: (303) 759–4349

Mountain Plains Branch
344 East Foothills Parkway, Suite #1EB
Fort Collins, CO 80525
Phone: (970) 207–0778
Toll-free: (888) 424–5071 (Northeastern
    CO & Southeastern WY)
Fax: (970) 207–0779

Southern Colorado Branch
2910 Beacon Street
Colorado Springs, CO 80907
Phone: (719) 520–5711
Phone: (719) 544–1136 (from Pueblo
    and Canon City)
Fax: (719) 447–0470

Connecticut
Southern New England Chapter
35 Cold Spring Road, Suite 411
Rocky Hill, CT 06067
Phone: (860) 563–1177
Toll-free: (800) 541–8350
Fax: (860) 563–6018
AFSNECT@AOL.COM

Delaware
Delaware Chapter
100 W. Tenth Street, Suite 206
Wilmington, DE 19801
Phone: (302) 777–1212
Toll-free: (800) 292–9599
Fax: (302) 777–1841
afdc@sprynet.com

Support Groups:
    (Call the chapter office for
    information)
Wilmington: Dupont Hospital for
    Children, 1600 Rockland Road,
    Room 1F51
Dover: Luthur Towers #1, 430 Kings
    Highway (Original Building)
Sussex: Nanticoke Hospital, 801
    MiddleFord Road, Stevens Room

District of Columbia
Metropolitan Washington Chapter
4455 Connecticut Avenue, N.W.,
    Suite 300
Washington, DC 20008
Phone: (202) 537–6800
Fax: (202) 537–6859
arthritsdc@aol.com

Florida
Florida Chapter
5211 Manatee Avenue, West
Bradenton, FL 34209
Phone: (941) 795–3010
Fax: (941) 798–3659
arthrpogfl@aol.com

Northwest Branch
1813 C West Fairfield Drive
Pensacola, FL. 32501
Phone: (850) 432–4348
Toll-free: (800) 578–7183
rsaurey@aol.com

Northeast Branch
318 Palmetto Street
Jacksonville, FL 32202
Phone: (904) 353–5770
Fax: (904) 353–7508
arth@southeast.net

Central Branch
2699 Lee Road, Suite 330
Winter Park, FL 32789
Phone: (407) 647–0045
Toll-free: (800) 510–5696
Fax: (407) 647–0064
arthritis@netpass.com

Gulfcoast Branch
9721 Executive Center Drive North,
   Suite 210
St. Petersburg, FL 33702
Phone: (813) 576–1727
Toll-free: (800) 850–9455
Fax: (813) 579–0628
Gulfcoast4@aol.com

Southwest Branch
6221 14th Street West, Suite 305
Bradenton, FL 34207
Phone: (941) 739–2729
Toll-free: (800) 741–4008
Fax: (941) 765–4168
arthritisw@aol.com

Mideast Branch
327 Okechobee Boulevard
West Palm Beach, FL 33401
Phone (561) 833–1133 (West Palm
   Beach)
(561) 278–2701 (Delray Beach)
Fax: (561) 835–0470
creakybone@aol.com

FM Support/ Groups
Palm Beach Gardens: Nativity Lutheran
   Church, 2075 Holly Drive, 746–1591
   or 746–6203.
Boca Raton: West Boca Community
   Outreach Center, Kimberly
   Boulevard & Lyons, 487–1162.
Loxahatchee: Palms West Hospital, 1st
   Floor Board Room, 798–8184
Port St. Lucie: First United Methodist
   Church, 290 S.W. Prima Vista
   Boulevard, Room 212, 335–4389
Boynton Beach Comprehensive Rehab,
   Inc., Gulfstream Mall (Gulfstream &
   Federal Highway) 3629 S. Federal
   Highway, 731–5094
Lake Worth: Lake Worth Rehab
   Association, 7173 Lake Worth Road
   (Worth Plaza), Lake Worth, 357–0010

Broward Branch
8201 North University Drive, Suite 201
Tamarac, FL 33321
Phone: (954) 726–6707
Fax: (954) 726–0118
broarth@icanect.net

Southeast Branch
4649 Ponce De Leon Boulevard,
Suite 305
Coral Gables, FL 33146
Phone: (305) 669–6870
Fax: (305) 669–6869
achyf@aol.com

Georgia
Georgia Chapter
550 Pharr Road, Suite 550
Atlanta, GA 30305
Phone: (404) 237–8771
Toll-free: (800) 933–7023 (Georgia only)
Fax: (404) 237–8153

Metropolitan Atlanta Branch
550 Pharr Road, Suite 550
Atlanta, GA 30305
Phone: (404) 237–7786
Fax: (404) 237–8253

Southeast Georgia Branch
511 Habersham Street
Savannah, GA 31401
Phone: (912) 234–7022
Toll-free: (800) 797–5604 (Georgia only)

Southwest Georgia Branch
307 15th Street, Suite 214
Columbus, Ga 31901
Phone: (706) 576–4086

Hawaii
Hawaii Chapter
45–114 Kamehameha Highway,
    Suite 500
Kaneohe, HI 96744
Phone: (808) 235–3636
Fax: (808) 235–0120

Maui Branch
J. Walter Cameron Center
95 Mahalani Street
Wailuku, HI 96793
Phone: (808) 244–0290
Fax: (808) 244–0290

Idaho
See Utah listing

Illinois
Greater Chicago Chapter
303 East Wacker Drive, Suite 300
Chicago, IL 60601
Phone: (312) 616–3470
Toll-free: (800) 735–0096 (Northeast IL
    and Northwest IN only)
Fax: (312) 616–9281
gccinfo@arthritis.org

Greater Illinois Chapter
2621 N. Knoxville
Peoria, IL 61604
Phone: (309) 682–6600
Fax: (309) 682–6732
ArthritisGIL@Flink.com

Alton Branch
Madison County
One Memorial Drive
Alton, IL 62002
Phone: (618) 463–7247

McLean County Branch
108 Boeykens Place, Suite 115
Normal, IL 61761
Phone: (309) 451–0785
Fax: (309) 454–5769
Caring/Sharing Support Group, 3rd
    Thurs.
Educational seminar on FM, 1st Thurs.

Champaign Branch
314 S. Neil
Champaign, IL 61820
Phone: (217) 398–7815

Decatur Branch
Macon County
1800 Lakeshore Drive, Room T 312
Decatur, IL 62525
Phone: (217) 422–1740

Effingham County Branch
Effingham County
1020 Pelican
Effingham, IL 52401
Phone: (217) 347–7395

Galesburg Branch
2621 N. Knoxville
Peoria, IL 61604
Phone: (309) 682–6600

Harrisburg Branch
714 S. Commercial #274
Harrisburg, IL 62946
Phone: (618) 252–0394

Kankakee Branch
St. Mary's Hospital
475 W. Merchant, Room 4025
Kankakee, IL 60901
Phone: (815) 937–2461

Lincoln Branch
Logan County
527 Pulaski
Lincoln, IL 62656
Phone: (217) 732–2923

Pontiac Branch
Saint James Hospital
610 E. Water Street
Pontiac, IL 61764
Phone: (815) 842–2828 (ext. 2237)

Quincy Branch
Adams County
Blessing at 14th
Quincy, IL 62301
Phone: (217) 228–3208
Fax: (217) 222–1803

FM Support Group in Quincy at
Blessing Hospital at 14th, Conference
Room C. Call the Adams County
Branch at 228–3208 or 964–2133.

Rockford Branch
Winnebago County
2400 Rockton Avenue, Suite 223
Rockford, IL 61103
Phone: (815) 971–6380

Support Groups:
    (Call Chapter for information)
Rockford: YMCA, Log Lodge 200 Y
    Boulevard
Rockford: First Presbyterian Church,
    406 N. Main Street.

Springfield Branch
Sangamon County
P.O. Box 1894
Springfield, IL 62705
Phone: (217) 523–2200
Fax: (217) 525–6467

Spring Valley Branch
1232 Linden
LaSalle, IL 61301
Phone: (815) 224–1455

Indiana
Indiana Chapter
8646 Guion Road
Indianapolis, IN 46268
Phone: (317) 879–0321
Fax: (317) 876–5608
afindy@in.net

Northern Region
P.O. Box 2559
Elkhart, IN 46515
Phone: (219) 294–8888
Fax: (219) 296–9339

Southern Region
700 North Weinbach Avenue,
    Suite 102
Evansville, IN 47711
Phone: (812) 474–1381(voice & fax)

Iowa
Iowa Chapter
2600 72nd Street, Suite D
Des Moines, IA 50322
Phone & Fax: (515) 278–2603

Eastern Iowa Office
Mercy Medical Center
701 10th St., SE Room 21–34
Cedar Rapids, IA 52403
Phone: (319) 369–4488
Fax: (319) 369–6848

Kansas
Kansas Chapter
1602 East Waterman
Wichita, KS 67211
Phone: (316) 263–0116
Toll-free: (800) 362–1108
Fax: (316) 263–3260

Kentucky
Kentucky Chapter
410 W. Chestnut Street, Suite 750
Louisville, KY 40202
Phone: (502) 585–1866
Toll-free: (800) 633–5335 (KY only)
Fax: (502) 585–1657

Bluegrass Branch
366 Waller Avenue, Suite 102
Lexington, KY 40504
Phone: (606) 276–1496
Fax: (606) 278–7551

Louisiana
Louisiana Chapter & Southwest Region
17050 Medical Center, Suite 300
Baton Rouge, LA 70816
Phone: (504) 751–7007
Toll-free: (800) 673–7508
Fax: (504) 751–0005

North Central Region
820 Jordan Street, Suite 240
Shreveport, LA 71101
Phone: (318) 221–2900 or
(318) 221–2904

Southeast Region
3445 N. Causeway Boulevard,
    Suite 700
Metairie, LA 70002
Phone: (504) 835–1221
Fax: (504) 834–8831

Maine
Maine Region Office
930 Brighton Avenue
Portland, ME 04102
Phone: (207) 773–0595
Fax: (207) 773–9735
arthritis-me@juno.com

Maryland
Maryland Chapter
1777 Reisterstown Road, Suite 150
Baltimore, MD 21208
Phone: (410) 602–0160
Toll-free: (800) 365–3811
Fax: (410) 602–0420

Lower Eastern Shore Branch
1505 Emerson Avenue
Salisbury, MD 21801
Phone: (410) 749–8509

Southern Maryland Branch
714 B & A Boulevard
Severna Park, MD 21146
Phone: (410) 544–5433

Support Groups:
Charlotte: Charlotte Hall Public Library,
    (301) 274–3536
Annapolis: Maryland Automobile
    Insurance Fund, 1750 Forest Drive,
    (410) 544–5433
Severna Park: Our Shepherd Lutheran,
    400 Benfield Road, (410) 544–5433

FM Self Help: Physiotherapy Associates
    at 901 Commerce Road, Annapolis
    (410) 544–5433; Physiotherapy
    Associates, 203 Hospital Drive, Glen
    Burnie (410) 544–5433; and Severna
    Park Physical Therapy, 844 Ritchie
    Highway, Severna Park, (410)
    544–0773

Western Maryland/Eastern West
    Virginia Region
22 South Market Street
Frederick, MD 21701
Phone: (301) 663–0303
Toll-free: (800) 750–9078
Fax: (301) 663–4295

Massachusetts
Massachusetts Chapter
29 Crafts Street
Newton, MA 02158
Phone: (617) 244–1800
Toll-free: (800) 766–9449
Fax: (617) 558–7686

Michigan
Michigan Chapter
17117 West 9 Mile Road, Suite 950
Southfield, MI 48075
Phone: (248) 424–9001
Toll-free: (800) 968–3030
Fax: (248) 424–9005

Support Groups:

Macomb County, Warren: St. John-Macomb Hospital, 12000 E. 12 Mile Road

Monroe County, Monroe: First Baptist Church, 1602 N. Custer Road

Oakland County: Ferndale, Ferndale Public Library, 222 East Nine Mile

Rochester Hills: Barclay Physical Therapy, 555 Barclay Circle, Suite 110

West Bloomfield: Beaumont West, 6900 Orchard Lake Road, Classroom B

For information on all above, call the chapter.

Ann Arbor Region
Senior Health Building
5361 McAuley Drive, 2435
P.O. Box 995
Ann Arbor, MI 48106
Phone: (313) 572–3224

Support Groups:

Ann Arbor: St. Joseph Mercy Hospital, Senior Health Building, Room 2440, 5361 McAuley Drive (two different groups, (313) 572–3224

Central Michigan Branch
241 East Saginaw Street, 402
East Lansing, MI 48823
Phone: (517) 332–9450

Support Groups:

Charlotte: Hayes Green Beech Hospital

Lansing: AF Branch Office Basement

Call (517) 332–9450 for information on above groups.

Midland County: Midland, Trinity Lutheran, (517) 835–1618 for information.

Mid-Michigan Branch
1000 Professional Drive
Flint, MI 48532
Phone: (810) 230–8290

Support Groups:

Flint: AF Branch, 230–8290

Flint: FM Daytime Support & Ed. Group, AF office

Laper: Laper Regional Hospital, 230–8290

Owosso: Owosso FM Support & Education Group (new group forming, call chapter for information)

Southeast Michigan Region
17117 West Nine Mile Road, Suite 950
Southfield, MI 48075
Phone: (243) 424–9001

Southwestern Michigan Branch
2490 S. Eleventh Street, Suite IV
Kalamazoo, MI 49009
Phone: (616) 353–9672 or
(616) 353–9736

Support Groups:

Gilmore: Gilmore Center for Health Education

Bronson: Methodist Hospital, 353–9672

West Michigan Branch
215 Sheldon, S.E. Suite A
Grand Rapids, MI 49503
Phone: (616) 774–8730

Support Group:
Blodgett Hospital, 774–8730

Minnesota
Minnesota Chapter
830 Transfer Road
St. Paul, MN 55114
Phone: (612) 644–4108
Toll-free: (800) 333–1380
Fax: (612) 644–4219
afmn@aol.com

Northeast Branch Office
325 S. Lake Avenue, Suite 5101
Duluth, MN 55802
Phone: (218) 727–4730
Fax: (218) 727–4743
arthritine@aol.com

Mississippi
Mississippi Chapter
350 North Mart Plaza
P.O. Box 9185
Jackson, MS 39286
Phone: (601) 362–6283
Toll-free: (800) 844–8400
Fax: (601) 362–6469
Call the chapter for information on the
eight active chapters in the state.

Missouri
Eastern Missouri Chapter
8390 Delmar Boulevard
St. Louis, MO 63124
Phone: (314) 991–9333
Fax: (314) 991–4020

Western Missouri/Greater Kansas City
    Chapter
1100 Pennsylvania Avenue, Suite 400
Kansas City, MO 64105
Phone: (816) 842–0335
Fax: (816) 842–2847
arthfdn@coop.crn.org

Montana
Montana Branch
1643 Lewis Avenue, Suite 3
Billings, MT 59102
Phone: (406) 245–0231
Fax: (406) 245–0397

Nebraska
Nebraska Chapter
7101 Newport Avenue, Suite 304
Omaha, NE 68152
Phone: (402) 572–3040
Toll-free: (800) 642–5292 (outside
    Omaha)
Fax: (402) 572–3048

Central Nebraska Branch Office
747 N. Burlington Avenue, Suite 312
Hastings, NE 68901
Phone: (402) 463–9766
Fax: (402) 463–9611

Lincoln Branch Office
5539 South 27th Street, Suite 108
Lincoln, NE 68512
Phone: (402) 421–3311
Fax: (402) 421–3843

Nevada
Las Vegas Branch
2660 South Rainbow Blvd.,
Suite B-102
Las Vegas, NV 89102
Phone: (702) 367–1626
Fax: (702) 367–6381

New Hampshire
New Hampshire Region Office
59 School Street
Concord, NH 03301
Phone: (603) 224–9322
Toll-free: (800) 639–2113
Fax: (603) 224–3778
arthritisnh@juno.com

New Jersey
New Jersey Chapter
200 Middlesex Turnpike
Iselin, NJ 08830
Phone: (732) 283–4300
Fax: (732) 283–4633

South Jersey Area Office
496 N. Kings Highway 211
Cherry Hill, NJ 08034
Phone: (609) 482–0600
Fax: (609) 482–7606

New Mexico
New Mexico Chapter
P.O. Box 8022
124 Alvarado S.E.
Albuquerque, NM 87108
Phone: (505) 265–1545
Toll-free: (800) 999–8022
Fax: (505) 265–1547
Contact the chapter for information on
    their FM Support Group and the
    FM Family Support Group.

New York

Central New York Chapter
5858 E. Molloy Road, Suite 123
Syracuse, NY 13211
Phone: (315) 455–8553
Fax: (315) 455–8714
afcny@atsny.com

Broome County Branch
161 Riverside Drive, Room MO2
Binghamton, NY 13905
Phone: (607) 798–8048

Genesee Valley Chapter
2423 Monroe Avenue
Rochester, NY 14618
Phone: (716) 271–2510
Fax: (716) 271–2764

Support Groups:
Wayne County, Ontario County, and
    Monroe County (Monroe—in coop-
    eration with FM Association of
    Rochester, NY)

Long Island Chapter
501 Walt Whitman Road
Melville, NY 11747
Phone: (516) 427–8272
Fax: (516) 427–3546
pmca@pb.net

Northeastern New York Chapter
1717 Central Avenue, Suite 105
Albany, NY 12205
Phone: (518) 456–1203
Toll-free: (800) 420–5554
Fax: (518) 869–3123
104743.2041@compuserve.com

New York Chapter
122 East 42nd Street, 18th Floor
New York NY 10168
Phone: (212) 984–8700
Fax: (212) 878–5960

Western New York Chapter
2440 Sheridan Drive
Tonawanda, NY 14150
Phone: (716) 837–8600 (in Buffalo)
Fax: (716) 837–8606
wnyaf@buffnet

North Carolina

Carolinas Chapter
7 Woodlawn Green, Suite 217
5019 Nations Crossing
Charlotte, NC 28217
Phone: (704) 529–5166
Toll-free: (800) 883–8806 (outside
    Charlotte)
Fax: (704) 529–0626

Central Region
3500 Vest Mill Rd, Suite 4
Winston-Salem, NC 27103
Phone: (910) 659–9776
Fax: (910) 659–7975

Eastern Region
3200 Beechleaf Court, Suite 100–21
Raleigh, NC 27604
Phone: (919) 873–0100
Fax: (919) 873–0108

Ohio

Central Ohio Chapter
3740 Ridge Mill Drive
P.O. Box 218182
Columbus, OH 43221
Phone: (614) 876–8200
Fax: (614) 876–8363

Coshocton County Branch
P.O. Box 1371
Coshocton, OH 43812
Phone: (614) 622–1202

Delaware County Branch
c/o Delaware Health Department
115 N. Sandusky Street
Delaware, OH 43015
Phone: (614) 368–1700

Eastern Ohio Branch
Eastern Ohio Regional Hospital
90 North 4th Street
Martins Ferry, OH 43935
Phone: (614) 633–4113

Fairfield County Branch
104 N. Broad Street, 3rd Floor
Lancaster, OH 43130
Phone: (614) 654–4334

Knox County Branch
915 E. High Street
Mt. Vernon, OH 43050
Phone: (614) 392–4528

Licking County Branch
YWCA Elder Services
126 West Church Street
Newark, OH 43055
Phone: (614) 349–3876

Muskingum Valley Branch
c/o The Masonic Temple
38 N. 4th Street, Suite 103
Zanesville, OH 43701
Phone: (614) 452–8144

Southeast Ohio Branch
Marietta Memorial Hospital
401 Matthews Street
Marietta, OH 45750
Phone: (614) 373–3083

Northeastern Ohio Chapter
Chagrin Plaza East
23811 Chagrin Boulevard 210
Cleveland, OH 44122
Phone: (216) 831–7000
Toll-free: (800) 245–2275 (outside
   Cuyahoga County)
Fax: (216) 831–1764
arthfdnneo@netincom.com

Akron/Canton Office
4767 Higbee Avenue, NW
Canton, OH 44718
Phone: (330) 492–0080
Toll-free: (800) 342–0746 (outside Stark
   County)

Fax: (330) 492–0266
Both the Cleveland and Canton
   Chapters have FM Self-Help
   Courses and FM Information
   Meetings. Contact the offices for
   more information.

Northwestern Ohio Chapter
309 North Reynolds Road, Suite F
Toledo, OH 43615
Phone: (419) 537–0888
Fax: (419) 537–6553

Ohio River Valley Chapter
7811 Laurel Avenue
Cincinnati, OH 45243
Phone: (513) 271–4545
Toll-free: (800) 383–6843
Fax: (513) 271–4703
aforvc@one.net

Dayton Branch
c/o KMC Rehabilitation Medical
   Department – C158
3535 Southern Building
Kettering, OH 45249
Phone: (513) 293–5211
Fax: (513) 293–5493

Scioto Valley Branch
902 Washington Street
(Serving West Virginia)
Portsmouth, OH 45662
Phone: (614) 353–4774
Toll-free: (800) 358–0380

Warren County Office
741 Senior Center, Room 223
570 North State Route 741
Lebanon, OH 45036
Phone: (513) 933–2263

Oklahoma
Oklahoma Chapter
2915 North Classen Boulevard,
   Suite 325
Oklahoma City, OK 73106
Phone: (405) 521–0066
Fax: (405) 521–0070

Eastern Oklahoma Chapter
4520 South Harvard, No. 100
Tulsa, OK 74135
Phone: (918) 743–4526
Toll-free (800) 400–4526
Fax: (918) 743–6910

Oregon
Oregon Chapter
4412 S.W. Barbur Boulevard,
    Suite 220
Portland, OR 97201
Phone: (503) 222–7246
Toll-free: (800) 283–3004
Fax: (503) 222–5542
arthore@teleport.com

Pennsylvania
Central Pennsylvania Chapter
17 South 19th Street
P.O. Box 668
Camp Hill, PA 17011
Phone: (717) 763–0900
Fax: (717) 763–0903

Lancaster/Lebanon Counties Branch
630 Janet Avenue
Lancaster, PA 17601
Phone: (717) 397–6271

Eastern Pennsylvania Chapter
Architects Building
117 South 17th Street, Suite 1905–15
Philadelphia, PA 19103
Phone: (215) 665–9200
Toll-free: (800) 355–9040 (in PA)
Fax: (215) 665–9249
esfida@arthritis.org

Allentown Branch
1227 Liberty Street, Suite 101–B
Allentown, PA 18102
Phone: (610) 776–6632
Fax: (610) 776–6654

Reading Office
401 Buttonwood Street
West Reading, PA 19611
Phone: (610) 375–0832

Wilkes-Barre Branch
Kirby Health Center
71 N. Franklin Street
Wilkes-Barre, PA 18701
Phone: (717) 823–2888
Fax: (717) 823–5038

Western Pennsylvania Chapter
Warner Center - 5th Floor
332 Fifth Avenue
Pittsburgh, PA 15222
Phone: (412) 566–1645
Toll-free: (800) 522–9900
Fax: (412) 391–1677

Erie Branch
12 E. Ninth Street, Room 304
Erie, PA 16501
Phone: (814) 455–0672

Support Groups: Contact the chapter
    for information on their two groups,
    one in Eire, and the other in
    Meadville.

Laurel Highlands Branch
1011 Old Salem Road
Greensburg, PA 15601
Phone: (412) 836–3370

Support Group:
Greensburg: The Healthplace,
    Westmoreland Mall, 832–9300

Rhode Island
Rhode Island Area Office
37 North Blossom Street
East Providence, RI 02914
Phone: (401) 434–5792
Fax: (401) 434–5779
AFSNECT@AOL.COM

South Carolina
Western Region
1754 Woodruff Road, Suite 107
Greenville, SC 29607
Phone: (864) 627–8833
Fax: (864) 627–8822

Tennessee
Tennessee Chapter
Midtown Plaza
1719 West End Avenue, Suite 303–W
Nashville, TN 37203
Phone: (615) 320–7626
Fax: (615) 320–7399

Greater Chattanooga Branch
735 Broad Street, Suite 203
Chattanooga, TN 37402
Phone: (423) 265–9244
Fax: (423) 265–9245

Nashville Branch
210 25th Avenue North, Suite 1100
Nashville, TN 37203
Phone: (615) 329–3431
Fax: (615) 321–0139

Smoky Mountain Branch
212 South Peters Road, Suite 105
Knoxville, TN 37923
Phone: (423) 470–7909
Toll-free: (800) 832–2061 (Tri Cities)
Fax: (423) 470–9356

West Tennessee Branch
5050 Poplar Avenue, Suite 1526
Memphis, TN 38157
Phone: (901) 685–9060
Fax: (901) 685–3277

Texas
North Texas Chapter
2824 Swiss Avenue
Dallas, TX 75204
Phone: (214) 826–4361
Toll-free: (800) 442–6653
Fax: (214) 824–5842

Heart of Texas Branch
6801 Sanger Avenue, Suite 220
Waco, TX 76710
Phone: (254) 772–9303
Fax: (254) 772–9937
hotarthritis@worldnet.att.net

Northwest Texas Chapter
3001 West 5th Street
Fort Worth, TX 76107
Phone: (817) 820–0635
Toll-free: (800) 283–7733
Fax: (817) 820–0642

South Texas Chapter
3701 Kirby Drive, Suite 1230
Houston, TX 77098
Phone: (713) 529–0800
Toll-free: (800) 364–8000
Fax: (713) 529–6622

Rio Grande Valley Branch
One Park Place, Suite 319
McAllen, TX 78503
Phone: (956) 630–0870
Toll-free: (800) 284–2483
Fax: (956) 630–3075

San Antonio Branch Office
8918 Tesoro Drive, Suite 590
San Antonio, TX 78217
Phone: (210) 829–7573
Toll-free: (800) 284–2483
Fax: (210) 829–4246

Utah
Utah/Idaho Chapter
448 East 400 South, Suite 103
Salt Lake City, UT 84111
Phone: (801) 536–0990
Toll-free: (800) 444–4993 (outside the
    Salt Lake area)
Fax: (801) 536–0991

Vermont
Vermont and Northern New York
    Region Office
P.O. Box 422
257 South Union Street
Burlington, VT 05401
Phone: (802) 864–4988
Toll-free: (800) 639–8838
Fax: (802) 864–5339
arthritisvtall@juno.com

Virginia

Virginia Chapter
3805 Cutshaw Avenue, Suite 200
Richmond, VA 23230
Phone: (804) 359–1700
Toll-free: (800) 456–4687 (Virginia only)
Fax: (804) 359–4900
vachapter@inetconn.net

Hampton Roads Branch
900 Commonwealth Place, Suite 101
Virginia Beach, VA 23464
Phone: (757) 420–4638
Phone: (757) 872–8848 (Peninsula)
Toll-free: (800) 456–4687
Fax: (757) 420–5078
arth-hrb@visi.net

Washington

Washington State Chapter
3876 Bridge Way North, Suite 300
Seattle, WA 98103
Phone: (206) 547–2707
Toll-free: (800) 542–0295
Fax: (206) 547–2805

North Puget Sound Branch
809 East Chestnut Street
Bellingham, WA 98225
Phone: (360) 733–2866
Toll-free: (800) 542–0295
Fax: (360) 671–5750

Southwest Washington Branch
1501 Pacific Avenue, Suite B01
Tacoma, WA 98402
Phone: (253) 274–8846
Fax: (253) 274–8847

Wisconsin

Wisconsin Chapter
8556 West National Avenue
West Allis, WI 53227
Phone: (414) 321–3933
Toll-free: (800) 242–9945
Fax: (414) 321–0365
afwc1@ibm.net

Call chapter office for information on support groups in the following communities:

Baraboo: St. Clare Hospital, 707 14th Street

Brookfield: Elmbrook Memorial Hospital, 19333 W. North Avenue

Burlington: The Rehab Center, Highway 36 North

Fond Du Lac: St. Agnes Hospital, 435 E. Division Street

Fort Atkinson: Fort Atkinson Memorial Health Services, 611 E. Sherman Avenue

Glendale: Laurel Oaks Retirement Center, 1700 W. Bender Road

Green Bay: Cerebral Palsy Center, 2801 S. Webster Avenue

Green Bay: Schneider National, 3101 S. Packerland Drive, Room LTRA

Greenfield: Southwest YMCA, 11311 W. Howard Avenue

Hartford: Hartford Memorial Hospital 1032 E. Sumner Street

Ironwood, MI: Grandview Hospital, Grandview Lane, Conference Room

Janesville: Mercy Clinic - East, 3524 N. Washington Street

La Crosse: Gundersen Lutheran, 1910 South Avenue, Overholt Auditorium

Lake Geneva: Mariner Outpatient Care, 350 Peller Road

Madison: Firstar Bank, 715 N. Midvale Boulevard

Madison: Parkside Presbyterian Church, 4002 Lein Road

Marshfield: Marshfield Clinic, 1000 N. Oak Avenue, Lower Level Classroom, #207

Medford: Taylor County Memorial Hospital

Milwaukee: Columbia Arthritis Center, 2025 E. Newport Avenue

Milwaukee: St. Joseph's Hospital, Mayfair Atrium Building, 10400 W. North Avenue

Milwaukee: St. Luke's Medical Center, 2900 W. Oklahoma Avenue

Monroe: Pleasantview Nursing Home, Highway 81

Oshkosh: Mercy Oakwood Medical Center, Presence Conference Room, 2700 W. 9th Avenue

Reedsburg: Reedsburg Area Medical Center, 2000 N. Dewey Street

Rhinelander: St. Mary's Hospital, 1044 Kabel Avenue

River Falls: River Falls Area Hospital, 422 N. Dallas Street

Stevens Point: Ruth Gilfrey Building, 817 Whiting Avenue,

Sturgeon Bay: Door County Library, 107 S. 4th Avenue

West Bend: St. Joseph's Community Hospital, 551 Silverbrook Drive

Wisconsin Rapids: Riverview Hospital, 410 Dewey Street - Conference Room D

Northeastern District Office
1253 South Irwin Avenue
Green Bay, WI 54301
Phone: (414) 432–5533
Southwestern District Office
802 West Broadway, Suite 206
Madison, WI 53713
Phone: (608) 221–9800
Fax: (608) 221–9696

OTHER HELPFUL RESOURCES

Because these large organizations have the ability to update material much faster than I, I urge you to contact them for information on local support groups and health care professionals. The Fibromyalgia Association of Greater Washington (FMAGW) has collected information nationwide for a North American Directory of Fibromyalgia Support Services that they sell for $16 for non-members and $13 for members. They also have a Support Group Leaders' Information Kit which costs $3. The Fibromyalgia Network also has information on support groups and healthcare professionals.

Informational materials available from these resources varies in cost and format so you should contact each one, requesting information on what they have available.

Inclusion of an organization or material in this listing is not an endorsement or agreement with anything published by these groups. Neither this book nor any information found at these sources is meant to take the place of medical treatment offered by a doctor.

Agency for Health Care Policy and Research Clearinghouse
Toll-free: (800) 358–9295

American Academy of Allergy and Asthma and Immunology
6113 E. Wells Street
Milwaukee, WI 53202
Phone: (414) 272–6071
Fax: (414) 276–6070
Toll-free: (800) 822–2762
info@aaaai.org
www.aaaai.org

American Academy for Environmental Medicine
P.O. Box 16106
Denver, CO 80216
Phone: (303) 622–9755

American Association for Chronic Fatigue Syndrome
7 Van Buren Street
Albany, NY 12206
Phone: (206) 521–1932

American Association of
Naturopathic Physicians
2366 Eastlake Avenue E. Suite 322
Seattle, WA 98102
Phone: (206) 328–8510
Fax: (206) 323–7612
74702.3715@compuserve.com

American Association of Retired
    Persons
National Headquarters
601 E. Street N.W.
Washington, D.C. 20049
Phone: (202) 434–2277
www.aarp.org.

American Chronic Pain Association
P.O. Box 850
Rocklin, CA 95677
Phone: (916) 632–0922
Fax: (916) 632–3208

American College of Rheumatology
60 Executive Park South, Suite 150
Atlanta, GA 30329
Phone: (404) 633–3777
Fax: (404) 633–1870
acr@rheumatology.org
www.rheumatology.org

American Disability Association
2121 8th Avenue, North, Suite 1623
Birmingham, AL 35203
Phone: (205) 323–3030
Fax: (205) 251–7417

American FMS Association, Inc.
6380 E. Tanque Verde Road,
    Suite D
Tucson, AZ 85715
Phone: (520) 733–1570
Fax: (520) 290–5550

American Lupus Society
260 Maple Court, 123
Ventura, CA 93003
Phone: (805) 339–0443
Fax (805) 339–0467
Toll-free: (800) 331–1802

American Occupational Therapy
    Association
1383 Piccard Drive
Rockville, MD 20850
Phone: (301) 948–9626

American Osteopathic Association
142 E. Ontario Street
Chicago, IL 60611
Phone: (312) 202–8000
Toll-free: (800) 621–1773
Fax: (312) 280–3860

American Physical Therapy Association
1111 N. Fairfax Street
Alexandria, VA 22314
Phone: (703) 684–2782

Americans With Disabilities Act
Office of Equal Employment
    Opportunity
Toll-free: (800) 366–6056

Arthritis Foundation Information
    Hotline
Toll-free: (800) 283–7800

Asthma and Allergy Foundation
    of America
1125 15th Street, NE, 502
Washington, DC 20005
Phone: (202) 466–7643
Toll-free: (800) 7–ASTHMA

British Columbia Fibromyalgia Society
Box 15455
Vancouver, British Columbia
V6B 5B2 Canada
Phone: (604) 430–6643
Fax: (604) 255–5836
    FM Forum
    Fibromyalgia: Face to Face

Centers for Disease Control
Fax Information Service
1600 Clifton Road NE
Atlanta, GA 30333
Phone: (404) 639–3534
Call (404) 332–4565 for instructions

Centers for Disease Control
    Information Line
Information Resources Management
    Office
Mail Stop C-15
1600 Clifton Rd., NE
Atlanta, GA 30333
(404) 332–4555

CFIDS Activation Network
P.O. Box 345
Larchmont, NY 10538
Phone: (212) 627–5631
Fax: (914) 636–6515

CFIDS Association of America Inc.
P.O. Box 220398
Charlotte, NC 28222–0398
(800) 442–3437
www.cfids.org
    The CFIDS Chronicle
    A Doctor's Guide to CFS by David
        Bell
    Hope and Help for CFS by Karyn
        Feiden
    Running On Empty: Chronic Fatigue
        Immune Dysfunction by Katrina
        Berne, Ph.D.

The CFIDS and Fibromyalgia Health
    Resource
1187 Coast Village Road, Suite 1-280
Santa Barbara, CA 93108–2794
Toll-free: (800) 366–6056
    Health Watch
    Alternative Medicine Yellow Pages

CFIDS Pathfinder
P.O. Box 2644
Kensington, M.D. 20891–2644
Phone: (301) 530–8624

Center for Mind-Body Medicine
5225 Connecticut Avenue, NW, 414
Washington, DC 20015
Phone: (202) 966–7338

Chronic Fatigue Syndrome
Fibromyalgia Syndrome Support
    Group
Bloomington-Normal, IL
Evonne B. Wumnest
303 Belview Avenue
Normal, IL 61761
Phone: (309) 452–2477 or
    (309) 452–9200

Chronic Pain Outreach Of
    Greater Saint Louis
P.O. Box 31686
Des Peres, MO 63131
Phone: (314) 768–1350
    Chronic Pain Letter
    Tired All the Time: How to Regain
        Your Lost Energy by Ronald
        Hoffman, M.D.

Clearinghouse on Disability
    Information
Office of Special Education and Rehab
    Services
U. S. Department of Education
Room 3132, Switzer Building
Washington, DC 20202
Phone: (202) 205–8241

Dallas Fibromyalgia Support Group
sponsored by Healthsouth Medical
    Center
2124 Research Row
Dallas, TX 75235
Phone: (972) 407–1985 or
    (972) 490–4926

Digestive Diseases Information
    Clearinghouse
National Digestive Diseases
    Information Clearinghouse (NDDIC)
2 Information Way
Bethesda, MD 20892–3570
Phone: (301) 654–3810
Fax: (301) 907–8906

Disability Rights Education and
Defense Fund
2212 Sixth St.
Berkeley, CA 94710
Phone: (501) 644–2555
ADA Hotline: (800) 466–4232
(9 a.m. to 5 p.m. PST)
Fax: (510) 841–8645

Documents on Health Programs
General Accounting Office (GAO)
P.O. Box 6015
Gaithersburg, MD 20884–6015
Phone: (202) 512–6000
Fax: (301) 258–4066
TDD: (301) 413–0006
www.infoww.gao.gov
> GAO reports on health, social secu-
> rity, welfare, veterans issues, and
> more

Endometriosis Association
8585 N. 76th Place
Milwaukee, WI 53223
Phone: (414) 355–2200
Toll-free: (800) 992–3636
Fax: (414) 355–6065

Fibromyalgia Alliance of America, Inc.
(FMAA)
P.O. Box 21990
Columbus, OH 43221–0990
Phone: (614) 457–4222
Fax: (614) 457–2729
> *Fibromyalgia Times*
> *Ohio '97: Fibromyalgia Research
> and Realities* videotape set of
> the conference, $30 per video (7
> in set) or $165 for the set
> *I have fibro what?* FMAA T-shirt, $15
> Membership $25 (includes quarterly
> newsletter)
> 10 percent discount for members

FM Association of Greater Washington
(FMAGW)
13203 Valley Drive
Woodbridge, VA 22191–1531
Phone: (703) 790–2324
Fax: (703) 494–4103
www.fmagw.org
> *Fibromyalgia Frontiers* newsletter.
> Membership $25 (1st year), renewal
> is $23/yr
> *North American Directory of
> Fibromyalgia Support Services*
> $16/13*
> *Support Group Leaders'
> Information Kit*
> plus an excellent selection of audio
> tapes of their speakers $10/8*.
> The Mid-Atlantic Conference on
> Fibromyalgia Treatment co-spon-
> sored by FMAGW and the
> Northern VA Institute for
> Continuing Medical Education in
> cooperation with National
> Institutes of Health $72/65*.
> *first price is for nonmembers and
> second is discount price for
> members.*
> *Contact FMAGW for information on
> complete listing and information
> on ordering.*

FM Association of Houston, Inc.
P.O. Box 2174
Bellaire, TX 77402
Phone: (713) 664–0180
> Videotapes:
> *Applying for Social Security* by
> Robert Hardy, JD
> *TMJ Disorder & FM* by C.R.
> Hoopingarner, DDS
> This group is in the process of writ-
> ing and publishing a book on
> fibromyalgia along the lines of
> "a practical guide to coping with
> FM" and are also in the process
> of updating thir videos. Contact
> them for an updated list of
> videos and other materials.

Fibromyalgia Association of Texas, Inc.
3810 Keele Drive
Garland, TX 75041
Phone: (972) 271–5085

Fibromyalgia Educational Systems, Inc.
500 Bushaway Road
Wayzata, MN 55391
Phone: (419) 843–3153
Fax: (419) 843–3155 or
(612) 473–6218 voice and fax
www.FMSedsys.com
    Taking Charge of Fibromyalgia
    A Patient Handbook & Educational
        Program
    A Self-Management Program for
        your Fibromyalgia Syndrome by
        Julie Kelly, MS, RN and Rosalie
        Devonshire, BA with contribu-
        tions by Jenny Fransen, RN, 4th
        ed., 1998

Fibromyalgia Management Association
Barbara Smith, Director
5101 Olsen Memorial Highway
Suite 2000
Golden Valley, MN 55422
Phone: (612) 827–1941

Fibromyalgia Network newsletter
    $19/yr US,$21/Canada, $24/outside
    North America;
Published by Health Information
Network, Inc.
P.O. Box 31750
Tucson, AZ 85751–1750
Toll-free: (800) 853–2929
Fax: (520) 290–5550
www.fmnetnews.com
Also: Advances in Research, Getting
    the Most Out of Your Medicines,
    FMS: A Patient's Guide, Support
    Group and Health Care Referrals
    also available

Hill-Burton Hospital Free Care
Toll-free: (800) 638–0742 or
(800) 492–0359

Interstitial Cystitis Association
P.O. Box 1553
Madison Square Station
New York, NY 10159
Phone: (212) 979–6057
Fax: (212) 677–6139

Job Accommodation Network
Toll-free (800) ADA–Work
(800) 526–7234 Voice TDD

LRH Publishing Co.
Box 100, Station A
Fredericton NB E3B 4Y2
Canada
    Quantity prices on request
    Booklets by Beth Ediger
    Coping With Fibromyalgia
    Fibromyalgia: Fight Back
    How to Run a Support Group: A
        Guide for Leaders
    Treating Fibromyalgia
    LRH Newsline

Lupus Foundation of America, Inc.
1300 Piccard Drive, Suite 200
Rockville, MD 20850
Phone: (301) 670–9292
Toll-free: (800) 558–0121
Fax: (212) 670–9486

Lyme Disease Foundation
1 Financial Plaza, 18th Floor
Hartford, CT 06103–2610
Phone: (860) 525–2000
Toll-free: (800) 886–LYME
Fax: (860) 525–8425

Massachusetts CFIDS Association
808 Main Street
Waltham, MA 02154
Phone: (617) 893–4415
Fax: (617) 227–5717
    CFIDS Update
    Chronic Fatigue Syndrome: A
        Primer for Physicians
    How to Apply for Social Security
        Disability Benefits If You Have
        CFS

The M. E. Association of Canada
246 Queen Street, Suite 400
Ottawa, Ontario K1P SE4, Canada
Phone: (613) 563–1565

MedicAlert Foundation
Toll-free: (800) 432–5378

Medical Rehabilitation Education
    Foundation
Toll-free: (800) GET–REHAB

National CFS and FMS Association
P.O. Box 18426
Kansas City, MO 64133
Phone: (816) 313–2000
    Heart of America Newsletter
    Social Security Disability Benefits
        Information

National Center for Homeopathy
801 N. Fairfax Street, Suite 306
Alexandria, VA 22314
Phone: (703) 548–7790

National Chronic Pain Outreach
    Association, Inc.
7979 Old Georgetown Rd.
Suite 100
Bethesda, MD 20814–2429
Phone: (301) 652–4948
Fax: (301) 907–0745

National Clearinghouse on Women
    and Girls with Disabilities
Educational Equity Concepts, Inc.
114 East 3rd Street
New York, NY 10016
Phone: (212) 725–1803

National FM Research Association, Inc.
P.O., Box 500
Salem, OR 97308
nfra@teleport.com
www.teleport.com/~nfra

National Headache Foundation
428 W. St. James Place, 2nd Floor
Chicago, IL 60614–2750
Phone: (312) 388–6399
Toll-free: (800) 843–2256
Fax: (312) 525–7357

National Institute of Allergy and
    Infectious Disease
9000 Rockville Pike,
Building 31 Room 7A32
Bethesda, MD 20892
Phone: (301) 496–5717
    Chronic Fatigue Syndrome other
        research reports

National Institute of Arthritis and
    Musculoskeletal and Skin Diseases
Arthritis Clearinghouse
1 AMS Circle
9000 Rockville Pike
Bethesda, MD 20892–3675
Phone: (301) 495–4484
Fax: (301) 587–2966
    Arthritis, Rheumatic Diseases, and
        Related Disorders Research
        Reports

National Health Information Center
Health Information Hotline
P.O. Box 1133
Washington, DC 20013–1133
Phone: (301) 565–4167
Toll-free: (800) 336–4797

National Institutes of Health
Communications Office
Building 10, Room 1C255
Bethesda, MD 20892
Phone: (301) 496–2563

National Institute of Environmental
    Health Sciences, Public Affairs
    Office
P.O. Box 12233
Research Triangle Park, NC 27709
Phone: (919) 541–3345

National Library of Medicine
8600 Rockville Pike
Bethesda, MD 20894
Phone: (301) 496–6308
Toll-free: (800) 272–4787

National Mental Health Association
Toll-free: (800) 969–6642

National Organization for Rare
    Diseases
P.O. Box 8923
New Fairfield, CT 06812–8923
Phone: (203) 746–6518
Toll-free: (800) 999–6673

National Rehabilitation Information
    Center
Toll-free: (800) 346–2742

National Self-Help Clearinghouse
25 West 43rd Street, Room 620
New York, NY 10036
Phone: (212) 642–2929
Fax: (212) 642–1956

National Organization of Social
    Security Claimants Representatives
(NOSCR – a lawyer referral source)
Toll-free: (800) 431–2804
    *Social Security Disability and SSI
        Claims: Your Need for
        Representation and Preparing
        for Your Social Security Disability
        of SSI Hearing*

National Women's Health Resource
    Center
2425 L Street, NW, 3rd Floor
Washington, D.C. 20037
Phone: (202) 293–6045
Fax: (202) 778–6306
    How Safe is Safe? by Doris Haire

Office on the Americans with
    Disabilities Act
Department of Justice, Civil Rights
    Division
P.O. Box 66798
Washington, D.C. 20035–6798
TTY: (800) 514–0383
Toll-free: (800) 514–0383

Office of Research on Women's Health
National Institutes of Health
Bethesda, MD
Phone: (301) 402–1770

Ontario Fibromyalgia Association
c/o The Arthritis Society
250 Bloor Street East, Suite 901
Toronto, ON M4W 3P2, Canada
Phone: (416) 967–1414
    Fibromyalgia: Face to Face

Sjogren's Syndrome Foundation, Inc.
333 N. Broadway
Jericho, NY 11753
Phone: (516) 933–6365
    The Sjogren's Syndrome Handbook
    Elaine K. Harris, ed.

Social Security Administration
Toll-free: (800) 772–1213 or local office
www.ssa.gov

Social Security Administration
Office of Disability
Professional Relations Branch
3–A–10 Operations
6401 Security Boulevard
Baltimore, MD 21235

Thyroid Foundation of America
Ruth Sleeper Hall, RSL 350–40
40 Parkman Avenue
Boston, MA 02114–2698
Phone: (617) 726–8500
Toll-free: (800) 832–8321
Fax: (617) 726–4136

To Your Health, Inc.
TyH Publications
(A Natural Goods Company)
17007 E. Colony Drive, Suite 105
Fountain Hills, AZ 85268
Toll-free: (800) 801–1406
Fax: (602) 837–1875
HealthPts1@AOL.com
   Health Points newsletter and
      catalog of products

U.S. Government Print Office
Superintendent of Documents
Washington, DC 20402
   American Rehabilitation

Vulvar Pain Foundation
P.O. Drawer 177
Graham, NC 27253
Phone: (910) 226–0704
Fax: (910) 226–8518

Well Spouse Foundation
610 Lexington Avenue, Suite 814
New York, NY 10022–6005
Phone: (212) 644–1241
Toll-free: (800) 838–0879
Fax: (212) 644–1338

PUBLICATIONS, TAPES, AND MORE

The following list includes newsletters, booklets, books, and audio and video tapes. Where possible I have included sufficient information to enable you to locate the source of such information. If no address is given, try your local bookstore, a large chain bookstore, or *www.Amazon.com* on the Internet. Prices vary. Amazon.com and other online bookstores usually give a discount but you still have to pay shipping and handling charges. Books sold through organizations may receive a discount.

I deleted a number of titles from the last edition, even though they are still good resources, such as Herbert Benson's *Maximum Mind*, because they are either out of print, available only in used condition, or through your library. Some books do not have a large distribution and may be available only through organizations.

This list is meant to be just a starting place. If you find something that you believe really helped you, send me a note in care of Taylor Publishing and I'll include it in the next edition.

Accent on Living Magazine
Accent Publications
P.O. Box 700
Bloomington, IL 61702–0700

The Aerobics Program for Total Well-
   Being: Exercise, Diet, Emotional
   Balance
by Kenneth Cooper, M.D., M.P.H.;
Bantam Books, New York, 1985.

Arthritis: A Take Care of Yourself
   Health Guide for Understanding
   Your Arthritis, 4th ed.
by James Fries;
Addison–Wesley Publishing Co., New York; 1995.

Herbert Benson, M.D.
Relaxation Response, 1976.
Beyond the Relaxation Response: How
   to Harness the Healing Power of
   Your Personal Beliefs, 1994.

Timeless Healing: The Power and
   Biology of Belief, audio cassette,
   1996.

The Wellness Book: The Comprehensive Guide to Maintaining Health and Treating Stress-Related Illness, 1993.

Betrayal by the Brain: The Neurologic Basis of Chronic Fatigue Syndrome, Fibromyalgia Syndrome, and Related Neural Network Disorders by Jay A. Goldstein, M.D., 1996.

A Companion Volume to Dr. Jay A. Goldstein's Betrayal by the Brain,
A Guide for Patients and Their Physicians
by Katie Courmel;
Haworth Medical Press
10 Alice Street,
Binghamton, NY 13904–1580, 1996.

Bridging the Gap: A National Directory of Services for Women and Girls With Disabilities, Educational Equity Concepts, New York, 1990.

Building Community: A Manual Exploring Issues of Women and Disability
Educational Equity Concepts
New York, 1985.

Deepak Chopra, M.D.
Ageless Body, Timeless Mind,
1995; *audio cassette,* 1993

Boundless Energy: The Complete Mind/Body Program for Overcoming Chronic Fatigue,
1997; *audio cassette,* 1993.

Creating Health, audio cassette, 1995.

Chronic Fatigue Syndrome & the Yeast Connection
by William G. Crook, M.D.
Professional Books, Inc.
Box 3246
Jackson, TN 38303; 1992.

Chronic Fatigue Syndromes: The Limbic Hypothesis by Jay A. Goldstein, M.D.
Haworth Medical Press
10 Alice Street
Binghamton, NY 13904
Toll-free: (800) 342–9678

Chronic Illness and Uncertainty: A Personal and Professional Guide to Poorly Understood Syndromes
by Don L. Goldenberg, M.D.
Dorset Press
P.O. Box 620026
Newton Lower Falls, MA 02162; 1996.

Clinical Overview and Pathogenesis of the Fibromyalgia Syndrome, Myofascial Pain Syndrome and Other Pain Syndromes
ed. I. Jon Russell, M.D., Ph.D.
Haworth Medical Press
10 Alice Street
Binghamton, NY 13904–1580; 1996.

Alex Comfort
New Joy of Sex, 1991.
The Joy of Sex Series, 1997.
Sexual Positions, 1997.
Sexual Foreplay, 1997.
Crown Publishers,
New York

Charles T. Kuntzleman and the editors of *Consumer Guide*
The Complete Book of Walking,
Pocket Books
NY, NY, 1992.

The Complete Drug Reference,
annual editions,
Consumer Reports
Toll-free: (800) 272–0722

Could Your Doctor Be Wrong?
by Jay A. Goldstein, M.D.
Pharos Publishing
New York, 1991.

Norman Cousins
Anatomy of an Illness as Perceived by the Patient, 1979.

Head First: The Biology of Hope and the Healing Power of the Human Spirit, 1990.

Diagnosing Your Doctor
Arthur R. Pell, Ph.D.
Chronimed Publishing, 1991.
Toll-free: (800) 848–2793

Directory of Pharmaceutical Indigent Assistance
The Pharmaceutical Research & Manufacturers of America
1100 15th St. NW
Washington, DC 20005
Phone: (202) 835–3400

Doctor, Why Am I So Tired?
Richard N. Podell, M.D., F.A.C.P.
Ballantine Books
New York, 1989.

Dr. Rosenfeld's Guide to Alternative Medicine: What Works, What Doesn't and What's Right for You
by Isadore Rosenfeld, M.D.
Random House
New York, 1996.

Examining Your Doctor: A Patient's Guide to Avoiding Harmful Medical Care
by Timothy B. McCall, M.D.
A Citadel Press Book
Carol Publishing Group
Secaucus, NJ, 1995.

Feeling Good: The New Mood Therapy
by David D. Burns, M.D.,1992.

Fibromyalgia & Chronic Myofascial Pain Syndrome, A Survival Manual
by Devin Starlanyl, M.D. and Mary Ellen Copeland, MS,MA
www.sover.net/~devstar
New Harbinger Publications, Inc.
5674 Shattuck Avenue
Oakland, CA 94609; 1996.

Fibromyalgia, Chronic Fatigue Syndrome, and Repetitive Strain Injury
Current Concepts in Diagnosis, Management, Disability, and Health Economics
ed. by Andrew Chalmers, M.D., F.R.C.P.C.A., Irving E. Salit, M.D. and Frederick Wolfe, M.D.
Haworth Medical Press
10 Alice Street
Binghamton, NY 13904–1580; 1995.

Fibromyalgia and You
ex. producer, I. Jon Russell, M.D.
Fibromyalgia Information Resources
P.O. Box 690402
San Antonio, TX 78269, *videotape*, 1995.

Fibromyalgia: A Comprehensive Approach
What You Can Do About Chronic Pain and Fatigue
by Miryam Ehrlich Williamson
Walker & Co.
435 Hudson Street
New York, NY 10014; 1996.

Fibromyalgia: Face to Face, videotape.
Ontario Fibromyalgia Association
250 Bloor Street East, Suite 901
Toronto, ON M4W 3P2 Canada

The Fibromyalgia Helpbook
by Jenny Fransen, RN & I. Jon Russell M.D., Ph.D.,
Smith House Press
St. Paul, MN; 1996.

Fibromyalgia Stretch Video
produced by Sharon Clark, Ph.D.
National Fibromyalgia Research Assoc.
P.O. Box 500
Salem, OR 97308

*The Fibromyalgia Syndrome: Current
    Research and Future Directions in
    Epidemiology, Pathogenesis and
    Treatment*
ed. by Stanley R. Pillemar, M.D.,
    Haworth Medical Press
    10 Alice Street
Binghamton, NY 13904–1580; 1994.

*The First Whole Rehab Catalog: A
    Comprehensive Guide to Products
    and Services for the Physically
    Disadvantaged*
by A. Jay Abrams & Margaret Ann
    Abrams
Betterway Publications, Inc.
White Hall, VA; 1990.

*For Convenience Sake: Practical
    Products for Easier Living*
409228 Howard Ave.
Kensington, MD 20895

For Your Special Needs
J. C. Penney Catalog
Health Care Products for the home
Toll-free: (800) 222–6161

From Fatigued to Fantastic!
by Jacob Teitelbaum, M.D.
Avery Publishing Group, 1996.

*Gardening for Handicapped & Elderly
    Persons*
bibliography by National Library Service
    for the Blind and Handicapped
Library of Congress, Washington, D.C.
    20542

Gastrointestinal Health
by Steven Peikin, M.D.
Harper Perennial
New York, 1991.

*Good Grief: A Constructive Approach
    to the Problem of Loss*
by Granger E. Westberg
Fortress Press
Philadelphia, PA; 1986.

Healing and the Mind
by Bill Moyers
Doubleday
New York, 1993.

*Healing the Body Betrayed: A Self-
    Paced, Self Guide to Regaining
    Psychological Control of Your
    Chronic Illness*
Robert Klein and Marcia Goodman, and
    D. Landau, Ph.D.s
Chronimed Publishing, 1992.
Toll-free: (800) 848–2793

*Healing Words: The Power of Prayer
    and the Practice of Medicine*
by Larry Dossey, M.D.
HarperSanFrancisco
HarperCollins, 1993.

Healthhouse USA
    catalog
P.O. Box 9036
Jericho, NY 11753
Phone: (516) 334–9754

*Hints & Tips to Make Life Easier:
    Practical Solutions for Everyday
    Problems*
Reader's Digest
Pleasantville, NY

How to Avoid Housework
by Paula Jhung
Fireside Books,
Simon & Schuster
New York, 1995.

How to Live Between Office Visits, 1994.
Love, Medicine & Miracles, 1990.
Peace, Love & Healing, 1990.
all by Bernie S. Siegel,M.D.
in several formats published by Harper
    Perennial
New York

In Sickness and in Health: Sex, Love, and Chronic Illness
by Lucille Carlton
Delacorte
New York, 1996.

It's Not All in Your Head
by Susan Swedo, M.D. & Henrietta Leonard, M.D.
HarperSanFrancisco
HarperCollins Publishing, 1996.

Job's Body: A Handbook for Bodywork
by Deane Juhan
Talman Company, 1991.

Job Strategies for People with Disabilities: Enable Yourself for Today's Job Market
by Melanie Astaire Witt
Peterson's Guides, 1992.

Journal of Musculoskeletal Medicine
55 Holly Hill Ln.
Box 4010
Greenwich, CT 06830
Phone: (203) 661–0600 or (212) 993–0440

Journal of Musculoskeletal Pain
I. Jon Russell, M.D., Ph.D., Editor
Haworth Medical Press
10 Alice St.
Binghamton, NY 13904
Toll-free: (800) 342–9678

Journal of the Chronic Fatigue Syndrome
Haworth Medical Press
10 Alice St.
Binghamton, NY 13904
Toll-free: (800) 342–9678

Living, Loving and Laughing with Pain
Jacqueline Jacobson Pliskin
Words & Pictures
P.O. Box 482
East Brunswick, NJ 08816; 1995.

Managing Pain Before It Manages You
by Margaret A. Caudill, M.D., Ph.D.
The Guilford Press
72 Spring Street
New York, NY 10012; 1994.

Manifesto for a New Medicine: Your Guide to Healing Partnerships and the Wise Use of Alternative Therapies
by James S. Gordon, M.D.
Addison-Wesley Publishing Co., New York; 1996.

Mastering Pain: A Twelve-Step Program for Coping with Chronic Pain,
by Richard A. Sternbach, director of Pain Treatment Center at Scripps Clinic & Research
Ballantine Publishing
New York, 1995.

Memory Minder: Personal Health Journal
Fibromyalgia Information Resources
P.O. Box 690402
San Antonio, TX 78269; 1996.

Mind As Healer—Mind as Slayer: A Holistic Approach to Preventing Stress Disorders
by Kenneth R. Pelletier
Delta Publishing
New York, 1992.

Musculoskeletal Pain Emanating from the Head and Neck: Current Concepts in Diagnosis, Management, and Cost Containment
ed. Murray Allen, M.D.
Haworth Medical Press
10 Alice Street
Binghamton, NY 13904–1580

*Musculoskeletal Pain, Myofascial Pain
Syndrome and the Fibromyalgia
Syndrome*
Haworth Medical Press
10 Alice Street
Binghamton, NY 13904
Toll-free: (800) 342–9678

*Myofascial Pain & Dysfunction: The
Trigger Point Manual*, Vol. 1 and
Vol. 2
by Janet Travell & David Simons
Williams & Wilkins, 1983.

*MYOPAIN '95
Abstracts from the 3rd World Congress
on Myofascial Pain & Fibromyalgia,
San Antonio, TX, USA*
ed. by I. Jon Russell, M.D., Ph.D.
Haworth Medical Press
10 Alice Street
Binghamton, NY 13904–1580

*New Our Bodies, Ourselves: A Book by
and for Women*
Touchstone Press, New York, 1996.

Older Women with Chronic Pain
by Karen A. Roberto, Ph.D.
Harrington Park Press
imprint of Haworth Press
10 Alice Street
Binghamton, NY 13904–1580, 1994.

Osler's Web
by Hillary Johnson
Crown Publishers
New York, 1996.

*The PDR Family Guide to Prescription
Drugs, 5th ed.*
Three Rivers Press, 1998.

*The PDR Pocket Guide to Prescription
Drugs,*
Pocket Books
New York, 1997.

*PDS Disability Facts
(and Trends Affecting Social Security
Disability Applicants)*
ed. by Douglas M. Smith, Esq. &
Barbara W. Smith
*PDS Disability Workbook for Social
Security Applicants*
By Douglas M. Smith, Attorney at Law
Physicians Disability Services, Inc.
P.O. Box 827
Arnold, MD 21012
Phone: (410) 974–1111
Fax: (410) 263–6636; 1995.

Mark Pellegrino, M.D.
*Fibromyalgia: Managing the Pain
The Fibromyalgia Survivor
The Fibromyalgia Supporter
Laughing at Your Muscles
Understanding Post-Traumatic
Fibromyalgia*
Anadem Publishing, Columbus, OH
43214,
Toll-free: (800) 633–0055

Post-Viral Fatigue Syndrome
Rachel Jenkins, MBBCHIR
James Mowbray, MBBCHIR
Haworth Medical Press
10 Alice Street
Binghamton, NY 13904
Toll-free: (800) 342–9678

Prescription for Anger
Gary Hankins, & Carol Hankins
Warner New York, 1994.

*Prescription for Nutritional Healing,
2nd ed.*, 1996.

*Prescription for Nutritional Healing:
A–Z Guide to Supplements*
by James F. Balch, M.D. & Phyllis
Balch,
CNC
Avery, 1998.

Bonnie Prudden
Myotherapy: Bonnie Prudden's Complete Guide to Pain-Free Living, 1985.

Pain Erasure: The Bonnie Prudden Way,
Ballantine Books
New York, 1985.

Sears Home Health Care
Sears Catalog
Toll-free: (800) 326–175

Sick and Tired of Feeling Sick and Tired: Living with Invisible Chronic Illness
by Paul J. Donoghue, Ph.D. and Mary Siegel, Ph.D.
W. W. Norton
New York, 1994.

Smart Patient, Good Medicine: Working with Your Doctor to Get the Best Medical Care
by R. L. Sribnick, M.D. and W. B. Sribnick, M.D.
Walker & Co.
435 Hudson St.
New York, NY 10014; 1994.

Streamlining Your Life: A 5-Point Plan for Uncomplicated Living
Stephanie Culp
Writers Digest Books
Columbus, OH, 1991.

Stretching
by Bob Anderson
Shelter Publications, 1987.

Successful Job Search Strategies for the Disabled: Understanding the ADA
by Jeffrey G. Allen
John Wiley & Sons
New York, NY, 1994.

TMJ, Its Many Faces
by Wesley E. Shankland, II, D.D.S., M.S.
Anadem Publishing, Inc.
Columbus, OH 43214; 1996.

Take Care of Yourself: The Complete Illustrated Guide to Medical Self-Care,
by Donald M. Vickery, M.D. & James F. Fries, M.D.
Addison-Wesley Publishing Co., New York; 1996.

Tired of Being Tired, Overcoming Chronic Fatigue & Low Energy
by Michael A. Schmidt
Frog, Ltd.
distributed by
North Atlantic Books
P.O. Box 12327
Berkeley, CA 94712; 1994.

The Truth About TMJ: How to Help Yourself
by Jennifer Hutchinson
Reinhardt & Still Publishers
Box 3232
Winchester, VA 22604
Toll-free: (800) 303–2244

Understanding Social Security and Disability Evaluation Under Social Security Administration
Baltimore, MD 21235
Toll-free: (800) 722–1213

We Laughed, We Cried: Life With Fibromyalgia
ed. by Kit Gardiser & Kathleen Kerry
KMK Associates
P.O. Box 60246
Palo Alto, CA 94306

*What Color is Your Parachute? A
   Practical Manual for Job-Hunters
   and Career Changers,* 1998.
*Job-Hunting Tips for the So-Called
   Handicapped People or People
   Who Have Disabilities: A
   Supplement to What Color Is Your
   Parachute?,* 1992.
by Richard Nelson Bolles,
   audio cassette, 1995.
Ten Speed Press, San Francisco, CA.

When You're Sick and Don't Know Why
by Linda Hanner & John Witek, M.D.
DCI Publishing
P.O. Box 47945
Minneapolis, MN 55447
Toll-free: (800) 848–2793

*Healing Wounded Doctor–Patient
   Relationships*
by Linda Hanner & Jon Witek, M.D.
Kashan Publishing
P.O. Box 307
Delano, MN 55328; 1995.

SURFING THE WORLD WIDE WEB

or How I Learned More Than I Ever Wanted To Know About FM On The Internet

Today, just three short years after the second edition of *When Muscle Pain Won't Go Away* was published, it is impossible to exclude a section on the Internet. While not everyone in the United States, much less the world, has access to a computer and therefore, the World Wide Web, it has already made a tremendous difference in how information is disseminated. When I researched the first two editions, I traveled to a medical library, searched by hand through heavy volumes of *Index Medicus*, then went upstairs to the stacks and tracked down each of the medical journals that contained the articles I needed. Then I spent a fortune at the photocopy machine making copies of the articles so I could carry them home and read and digest them at my own pace. I also spent quite a bit of money on purchasing books and tapes, as well as subscribing to the major FM newsletters and *Journal of Musculoskeletal Pain.*

For this edition, I have done the majority of my research from my desktop computer or from the laptop while

I was in bed. (I know many experts emphasize that you should do nothing in the bedroom but sleep; sounds great, but there are times that if I don't work in bed, I don't work. Sorry, docs, but a woman's gotta do what a woman's gotta do.)

If you want to find the most recent information available and you don't want to have to wait until it is gathered, analyzed, rewritten, and published into either a book or newsletter, turn to the World Wide Web or the Internet. This section gives you some tips, guidelines, and warnings about what you will find when you seek information on fibromyalgia online.

Web sites frequently change their URL or address, disappear, or are never updated. You have to sift through a great deal of garbage to find good, solid, accurate information because, no one controls the content of web sites. Be very careful about the information and advice that you find on fibromyalgia (or any health condition, as far as that goes). There is some excellent information available for someone newly diagnosed with fibromyalgia. You just have to sort it out from the rest. Unfortunately, there are too many people (and that includes some health professionals) who are more interested in getting your money than in improving your health.

Now that I've said that, let's take a look at how you can find information on fibromyalgia. We'll assume you are already subscribing to an online service, and that you are familiar with basic computer use. Commercial providers such as CompuServe and America On-Line provide a lot of service for their customers including direct access to the Internet itself. The average cost for most commercial services for unlimited monthly access is usually around $20.

Usually, the opening screen or main menu for a commercial service, contains a table of contents listing major topics available through the service. Choose health. You will see a menu or listing of databases, some of which are free and others which charge either on a connection basis or on a per search basis. Databases that do not have a surcharge include the HealthNet Medical Reference, which covers disorders and diseases, symptoms, drugs, surgeries/tests/procedures, home care, first aid, and more. I found it okay for a preliminary search but if I want more depth, I had to go elsewhere. On CompuServe's Health Database plus, the charge is $1.50 for each portion of a medical journal read or retrieved; for the Comprehensive Core Medical Library the charge is $3 for up to 5 titles and $3 for each group of 5 titles after that plus $3 for a full reference.

To access more health information from a wider range of sources, you may want to search the Web directly. Your service provider will provide direct access to the Internet where you will use what is called a search engine to find web sites that cover the topic in which you are intereted. The most common search engines are Yahoo!, Excite, HotBot, WebCrawler, Lycos, InfoSeek, and Snap but there are others and new ones are constantly being developed. You might want to set some time aside and try out the different search engines until you find one that suits your needs best.

Most will link you up with general health sites that provide information ranging from alternative medicine to information on drugs. When you search for a site on fibromyalgia, don't be surprised if you end up with nearly 2,000 sites. Some are duplicates, others are no longer being maintained. Search engines are designed to seek out the sites, but they cannot determine which ones are still active, nor will search engines weed out any duplicate sites which show up.

One word of warning: the Internet can become addicting. You set up a search and then you follow one link after another until suddenly it's 2:42 a.m and you realize you've just spent five hours "surfing the web."

You may suffer from information overload. Researching on the Internet is not about finding enough information but about sorting through it, tossing out what is extraneous, outdated, or just plain nonsense, and then organizing it into some sort of format so you can then learn how it applies to you and your fibromyalgia.

There are quite a number of electronic bulletin boards and chat rooms out there that can be a great deal of help to someone who's just been diagnosed with FM and doesn't have a local support group. They are very good for letting you know that you aren't alone in this and that you aren't crazy, but just as the Internet itself can be addicting, so can these electronic support systems. The first thing you know, you're spending all of your time on the computer and none in "real life." Just as a live support group must remain positive and allow you to do a little spouting off when things get really rough, so must an electronic, computer support group. The computer, the Internet, and all of the resources on it are a tool; use them wisely to help you learn more about fibromyalgia and to cope with a chronic illness.

I suggest the following sites as places for you to start, places where I know you will find quality information. It's up to you then to decide just how much time you can or are willing to spend searching, and just what information you really want. When you enter the web address, type http:// first, the address (or URL). Type the address just as it appears; any errors will keep you from hitting the site.

American Academy of Physical Medicine and Rehabilitation (AAPM&R) www.aapmr.org

American College of Rheumatology www.rheumatology.org

The Americans with Disabilities Act (ADA) www.usdoj.gov/crt/ada/

The American Medical Association (AMA) www.ama-assn.org/

American Occupational Therapy Association www.aota.org

Arthritis Foundation National Office www.arthritis.org

Centers for Disease Control & Prevention (CDC) www.cdc.gov/

Clinical Trials www.centerwatch.com

Colorado Health Net (NET) www.Coloradohealthnet.org

Combined Health Information Database (CHID) http://www.chid.nih.gov

Federal Register Online (official publication for all notices, rules and proposed rules from federal agencies and organizations, presidential documents and executive orders) www.access.gpo.gov/su_docs/aces/aces140.html

Federal Web Locator (links to federal agencies and organizations) www.law.vill.edu/Fed-Agency/feb-webloc.html

Fibromyalgia Newsgroup news:alt.med.fibromyalgia

Gulf War Veteran Resource Pages www.gulfwar.org

Healthfinder(trademark) (a gateway consumer health information web site from the U.S. government www.healthfinder.gov

HealthWeb: Rheumatology www.medlib.iupui.edu/hw/rheuma/home.html

Lupus Foundation of America www.internet-plaza.net/lupus/

Missouri Rehabilitation Research and Training Center (MAARTC) www.hsc.missouri.edu/arthritis

National Digestive Diseases Information Clearinghouse www.niddk.nih.gov/health/digest/digest.htm

National Fibromyalgia Research Association (NFRA) www.teleport.com/~nfra/

National Institute of Arthritis and Musculoskeletal and Skin Diseases (NIAMS) www.nih.gov/niams/

National Kidney and Urological Diseases Information Clearinghouse www.niddk.nih.gov/

National Library of Medicine (MEDLINE & other databases) www.nlm.nih.gov/databases

North American Chronic Pain Association of Canada (NACPAC) www.sympatico.ca/nacpac/

Oregon Fibromyalgia Foundation (OFF) Dr. Robert Bennett and Dr. Sharon Clark www.myalgia.com

Dr. Mark J. Pellegrino www.chronic-painsolutions.com/pellerin298.htm

Sapient Health Network (SHN) www.shn.net

Social Security Administration www.ssa.gov

Vulvodynia
  *www.med.umich.edu/obgyn/vulva/v
  ulvod.html*
Miryam E. Williamson's Fibromyalgia:
  Comprehensive Approach
  www.shaysnet.com/~wmson/

The list could go on, with more than enough information to fill a book in itself. There are several good books out on researching the web in general and on health subjects in particular, but they go out of date so quickly that I'm not comfortable in recommending them. If you feel that you do need written guidelines on online health searches, try to find one printed within the last year. It takes almost a year for the majority of books to go from completed manuscript to printed and bound book, available at the bookstore. When you are talking about the technology involved with computers and the Internet, that makes a book almost obsolete the moment you buy it.

You can't go wrong when you're dealing with the sites I've listed above. In fact, I can almost guarantee you will have more information than you will know what to do with. Just be aware of when it was updated last. When good news about the latest research on FM, CFS, and the other conditions we've discussed is released, you'll probably read all about it—online. Good luck.

PARTIAL LIST OF FIBROMYALGIA RESEARCHERS

This is just a partial list of individuals and institutions involved in research on fibromyalgia. More are being published in medical journals every year. For information on these physicians' policies for accepting new patients, contact the rheumatology or psychiatric clinic or the office of the individual listed. Most physicians are affiliated with a school of medicine but they may also have a separate practice at a clinic or hospital; this information is also listed if applicable.

If you want to participate in a clinical trial, you may have to stop all medications prior to the study and this may lead to an increase in your symptoms. However, you may also find that the trial's medications or treatment may be of benefit to you. Information gathering studies will not require you to change treatment.

Some researchers can be contacted through FM Associations, such as the Fibromyalgia Association of Greater Washington (FMAGW), the Fibromyalgia Alliance of America or the National Fibromyalgia Research Association or the Fibromyalgia Network, all of which are listed in the resource section. I have also indicated several researchers who accept donations for their projects.

There is a very real need for more funds for research into FM and the associated conditions. Although there has been some progress in obtaining government funding, fibromyalgia is severely underfunded in comparison to other chronic conditions when considering the number of individuals involved, lost income, and higher health care costs. The Arthritis Foundation also lists facilities that receive funding for arthritis research and that are currently conducting research on fibromyalgia. Check some of the web sites listed in the section on the Internet for more information. Information provided below on specific studies and guidelines for making donations to the researchers was provided by the FMAGW newsletter, *Fibromyalgia Frontiers* (Summer 1997). FMAGW also can provide a current list of researchers who accept direct donations from the public.

Alabama

Dennis W. Boulware, M.D.
Division of Rheumatology
University of Alabama-Birmingham
UAB Station VAH L205
Birmingham, AL 35294

Laurence A. Bradley, Ph.D.
Division of Clinical Immunology and
    Rheumatology
University of Alabama-Birmingham
    Station
475 Boshell Diabetes Building
Birmingham, AL 35294

Recent and/or ongoing studies:
1. Fibromyalgia: Central Factors in Its
   Etiopathogenesis (Studies of the
   Relationships Among Abnormal
   Pain Perception, Brain Blood Flow,
   and Substance P and Other
   Neuropeptides.)
2. Inhibited Functional Brain Activity in
   Fibromyalgia (Studies of Abnormal
   Brain Blood Flow Responses to
   Acute Pain and Other
   Environmental Stimuli). Time frame:
   Present to December 2000.

Note: Make checks payable to
    University Controller. Make notation
    on check that funds are "unrestrict-
    ed donation to Dr. Laurence A.
    Bradley's fibromyalgia research pro-
    gram."

California

P. Kahler Hench, M.D. F.A.C.P.
Senior Consultant
Division of Rheumatology
Scripps Clinic and Research
    Foundation
10666 North Torrey Pines Road
La Jolla, CA 92037

Daniel J. Wallace, M.D.
Associate Clinical Professor of
    Medicine
Department of Medicine
Division of Rheumatology
Cedars-Sinai Medical Center
UCLA School of Medicine
Los Angeles, CA 90048

Xavier J. Caro, M.D.
Northridge Medical Center Plaza
18350 Roscoe Boulevard Suite 418
Northridge, CA 91325

District of Columbia

Daniel J. Clauw, M.D.
Georgetown University Medical Center
3800 Reservoir Road, N.W., LL Gorman
Washington, DC 20007

Study topic:
Central Nervous System Dysregulation
    in Fibromyalgia, Chronic Fatigue
    Syndrome, and Related Disorders
    Time frame: 1995 – 2000
Note: Make check payable to Division
    of Rheumatology, Georgetown
    University Medical Center.

Illinois

Muhammad B. Yunus, M.D.
Assistant Professor of Medicine
Division of Rheumatology
Department of Medicine
University of Illinois
College of Medicine at Peoria
P.O. Box 1649
Peoria, IL 61656

Study topics:
1. Genetic Studies, including Twin
   Studies, in Fibromyalgia Syndrome.
   Time frame: 2 years.
2. 2 Positron Emission Tomography
   (PET) of the Brain in Fibromyalgia
   and Control Groups.
   Time frame: 18 months.

3. Relationship Between Irritable Bowel Syndrome and Fibromyalgia Syndrome.
   Time frame: 2 years.
Note: Make check payable to University of Illinois. Include notation on check that it is a donation for fibromyalgia research.

Howard M. Kravitz, D.O. M.P.H.
Department of Psychiatry
Rush Medical College
Rush-Presbyterian
St. Luke's Medical Center
1653 West Congress Parkway
Chicago, IL 60612

Alfonse Masi, M.D., D.R., P.H., F.A.C.P.
Professor of Medicine
University of Illinois
P.O. Box 1649
College of Medicine at Peoria
One Illini Drive
Peoria, IL 61656

Kansas
Frederick Wolfe, M.D.
Clinical Professor of Internal Medicine & Family & Community Medicine
University of Kansas School of Medicine and Arthritis Research Center
1035 North Emporia
Suite 230
Wichita, KS 67214

Maryland
Stanley R. Pillemer, M.D.
NIDR
Building 10, Center Drive M.S.C. 1190
National Institutes of Health
Bethesda, MD 20892

Robert D. Gerwin, M.D.
Assistant Professor of Neurology
Johns Hopkins University
7500 Hanover Parkway
Suite 201
Greenbelt, MD 20770

Massachusetts
Gail K. Adler, M.D., Ph.D.
Brigham & Women's Hospital
Endocrine Hypertension Division
221 Longwood Avenue
Boston, MA 02115

Study topics:
1. Regulation of the Hypothalamic-Pituitary Adrenal Axis in Fibromyalgia.
   Time frame: 3 years.
2. Effect of the Hypothalamic-Pituitary-Adrenal Axis on Sympathoadrenal Function in FMS.
   Time frame: 3 years.
Note: Make check payable to: Brigham & Women's Hospital. Make notation on check that it is for fibromyalgia studies conducted by Dr. Gail K. Adler.

Don L. Goldenberg, M.D.
Chief of Rheumatology,
Newton-Wellesley Hospital
Professor of Medicine
Tufts University School of Medicine
2000 Washington Street Suite 304
Newton, MA 02162

David T. Felson, M.D., MPH
Arthritis Center
Associate Professor of Medicine
Boston University School of Medicine
80 East Concord Street
A203
Boston, MA 02118

James Hudson, M.D.
Associate Professor of Psychiatry
Harvard Medical School
Boston, Massachusetts
Associate Chief
Biological Psychiatry Laboratory
Laboratories for Psychiatric Research and Psychosis Program
McLean Hospital
Belmont, MA 02178

Harrison G. Pope, Jr., M.D.
Associate Professor of Psychiatry
Harvard Medical School
Boston, Massachusetts
Chief, Biological Psychiatry Laboratory
Laboratories for Psychiatric Research
   and Psychosis Program
McLean Hospital
Belmont, MA 02178

Robert W. Simms, M.D.
Arthritis Center
Boston University School of Medicine
71 East Concord Street
Boston, MA 02118

Michigan
Jan E. Bachman, Ph.D.
Staff Psychologist,
Department of Anesthesiology
University of Michigan Medical Center
Ann Arbor, MI 48109

Leslie J. Crofford, M.D.
Assistant Professor
Internal Medicine/Rheumatology
University of Michigan
1150 W. Med Center Drive
Room 5510E MSR BI
Ann Arbor, MI 48109–0680

Randy S. Roth, Ph.D.
Lecturer
Departments of Anesthesiology &
   Physical Medicine & Rehabilitation
University of Michigan Medical Center
C233 Med Inn Building
Ann Arbor, MI 48109

Minnesota
Jeffrey M. Thompson, M.D.
Department of Physical Medicine and
   Rehabilitation
Mayo Clinic
Rochester, MN 55901

Missouri
John M. Uveges, Ph.D.
Psychology Service
University of Missouri
Columbia School of Medicine
Columbia, Missouri
Psychology Service
Harry S. Truman Memorial Veterans
   Hospital
800 Hospital Drive
Columbia, MO 65201

Susan P. Buckelew, Ph.D.
Associate Professor
Division of Clinical Health
Psychology & Neuropsychology
Department of Physical Medicine and
   Rehabilitation Center
One Hospital Drive
Columbia, MO 65201

New York
Arthur Weinstein, M.D.
New York Medical College
Division of Rheumatic Diseases &
   Immunology
Room G, 73 Munger Pavilion
Valhalla, NY 10595

Study topic:
Pathogenesis of Lyme-Induced
   Fibromyalgia.
   Time frame: 1995–1998
Note: Make check payable to:
   Research, Division of Rheumatic
   Disease and Immunology.

North Carolina
Daniel Hamaty, M.D.
Demas Neurology and
Medical Rehabilitation PA
2115 East 7th Street
Charlotte, NC 28204

Glen A. McCain, M.D., F.R.C.P.
Medical Director
Chronic Pain Service
Charlotte Institute of Rehabilitation
Carolinas Medical Center &
Associate Director
The Pain Therapy Center
Presbyterian Hospital
Southeast Arthritis Care Center
1521 East Third St.
Charlotte, NC 28204

Oregon
Robert M. Bennett, M.D., F.R.C.P.
Professor of Medicine
Chairman, Division of Arthritis &
    Rheumatic Diseases
Department of Medicine
Oregon Health Sciences
3181 SW Sam Jackson Park Road
Portland, OR 97201

Stephen M. Campbell, M.D.
Associate Professor of Medicine
Division of Arthritis and Rheumatic
    Diseases
The Oregon Health Sciences University
    Chief Rheumatology Section
Portland Veterans Administration
Medical Center
Portland, OR 97207

Carol S. Buckhardt, Ph.D.
Associate Professor of Nursing &
    Assistant Professor of Medicine
    (Research)
Division of Arthritis & Rheumatic
    Diseases
School of Nursing
Oregon Health Sciences
3181 SW Sam Jackson Park Road
Portland, OR 97201–3098

Sharon R. Clark, Ph.D.
Associate Professor of Nursing &
    Assistant Professor of Medicine
    (Research)
Division of Arthritis and Rheumatic
    Diseases
Oregon Health Sciences
3181 SW Sam Jackson L329A
Portland, OR 97201

Pennsylvania
Dennis C. Turk, Ph.D.
Professor of Psychiatry and
    Anesthesiology
University of Pittsburgh
Director, Pain Evaluation and
    Treatment Institute
Baum Boulevard at Craig Street
Pittsburgh, PA

Tennessee
Theodore Pincus, M.D.
Division of Rheumatology and
    Immunology
Medical Center
Vanderbilt University School of
    Medicine
203 Oxford House Box 5
Nashville, TN 37232–4500

Texas
Bernard R. Rubin, D.O., F.A.C.P.
Professor of Medicine
Chief, Section of Rheumatology
Department of Internal Medicine
University of North Texas
Health Science Center
3500 Camp Bowie Boulevard
Fort Worth, TX 76107–2699

I. Jon Russell, M.D., Ph.D.
Associate Professor of Medicine
Medicine/Clinical of Immunology
University of Texas Health Science
Center at San Antonio
7703 Floyd Curl Drive
San Antonio, TX 78284–7868

Study topics:
1. Fibromyalgia Syndrome: Abnormal Levels of Pain—Neurotransmitters in Cerebrospinal Fluid.
   Time frame: Ongoing.
2. Fibromyalgia Syndrome: Abnormal Platelet Function and Low Levels of Platelet Serotonin as a Marker for Defective Regulation of Neurotransmitters.
   Time Frame: Ongoing.
Note: Make check payable to University of Texas Health Science Center. Include notation on check: "FMS Research Monies—No Overhead Allowed."

Washington
Martin J. Kushmerick, M.D., Ph.D.
Department of Radiology, SB05
University of Washington
Seattle, WA

Canada
Roman Jaeschke, M.D.
St. Joseph's Hospital
Fontbonne Building
50 Charlton Avenue East
Hamilton, Ontario

Franklin A. Lue, M.S.C.. Deng, C.C.E.
Assistant Professor
Department of Psychiatry,
University of Toronto Centre for Sleep and Chronobiology
Western Toronto Hospital
399 Bathurst Street
Toronto, Ontario M5T 258

H. Mersky, M.D.
Department of Research
London Psychiatric Hospital
850 Highbury Avenue
P.O. Box 2532
London, Ontario

Eldon Tunks, M.D., F.R.C.P. (C)
Professor of Psychiatry
Chedoke-McMaster Pain Clinic
McMaster University
Hamilton, Ontario

Kevin P. White, M.D., F.R.C.P. (C)
Clinical Research Fellow
Rheumatology
University Hospital
University of Western Ontario
London, Ontario

Harold Merskey, D.M.
Professor of Psychiatry
University of Western Ontario
Director of Research
London Psychiatry Hospital
London, Ontario

Harvey Moldofsky, M.D.
Professor Department of Psychiatry and Medicine
University of Toronto
Toronto Western Hospital
399 Bathurst Street
Toronto, Ontario M5T 258

W.J. Reynolds, M.D., F.R.C.P.,
Rheumatology 1–204 FP
The Toronto Hospital, Western Division
399 Bathurst Street
Toronto, Ontario M5T 2S8

Hugh Smythe, M.D.
Professor and Head
Rheumatic Diseases Unit
University of Toronto
Toronto Hospital West Division
399 Bathurst Street, Fell Pav. 1–225
Toronto, Ontario M5T 2S8

ACTIVISM: RAISING AWARENESS & FUNDING

Webster's New World Dictionary defines activism as taking direct action to achieve a political or social end. In a perfect world, individuals with a chronic condition, particularly one as painful as fibromyalgia, should only have to worry about coping with their condition. We who must live with FM and everything that that diagnosis entails, know that the world is far from perfect. Although researchers are learning more about FM every day, we still do not know the exact cause nor do we have treatments that will completely or significantly reduce the symptoms of it. And it takes money to do research to achieve those goals.

Even though fibromyalgia affects more people than either chronic fatigue syndrome or rheumatoid arthritis, fibromyalgia receives less than either condition in funding from the National Institutes of Health for research. According to an article published in the *Fibromyalgia Times* (Summer 1997), a quarterly newsletter for the Fibromyalgia Alliance of America, $9 was spent on research at NIH for every person with CFS, $9.50 for every person with rheumatoid arthritis, while only 64 cents was spent for each person with FM. Yet the yearly health care cost for an individual is estimated to be $2,274 while it has been estimated that FM costs approximately $20 billion total in health care and lost work.

People who have the power to grant money for research must be made aware of a problem before steps can be taken to correct that problem. For years, progress in research on fibromyalgia was hampered by disbelief in FM as a legitimate diagnosis and, although advances have been made, we still have a long way to go. The United States government is the primary source of funding for research, and much of that goes through the National Institutes of Health in Bethesda, Maryland. The competition for the funds that are awarded for research from the NIH is very stiff, with approximately 25,000 applications for research received each year. Applications must go through a complex peer review process to determine who receives the funds.

For several years there have been a number of individuals who have been very active in increasing fibromyalgia awareness in NIH, its review committees, and Congress. There was an initial funding of $1.4 million earmarked for FM research in 1994, but the full amount was never used. In September 1996, Dr. Stephen Katz, Director of the National Institute of Arthritis and Musculoskeletal and Skin Diseases (NIAMS), met with ten patient representatives from FM organizations to discuss fibromyalgia. One problem discussed by these patient representatives was that too often those who review the proposals on FM have no knowledge of or expertise with fibromyalgia. One outcome of that meeting was to add fibromyalgia to the chronic fatigue syndrome Special Emphasis Panel (SEP). In a Special Emphasis Panel, peer reviewers who evaluate research proposals have knowledge in that particular field. A concern addressed in *Fibromyalgia Times* was that, by combining CFS and FM, one condition would lose funding. That will not be the case, because chronic fatigue syndrome is under the National Institute of Allergy and Infectious Diseases (NIAID) while FM remains under the National Institute of Arthritis and Musculoskeletal and Skin Diseases (NIAMS), which have separate budgets.

In September 1996, the Fibromyalgia Association of Greater Washington, Inc. (FMAGW), and the Northern Virginia Institute for Continuing Medical Education co-sponsored the Mid-Atlantic Conference on Fibromyalgia Treatment, which was presented in cooperation with the NIAMS. All believed the conference was a success because it increased the knowledge of fibromyalgia to a large group of health care professionals. (Audio tapes of the conference are available from FMAGW). This conference was an important step in gaining research credibility for FM and NIAMS involvement was significant.

However, since these beginnings there has not been a truly significant increase in money directed toward FM research. An article published in *Fibromyalgia Times* (Summer 1997) urged individuals to write to their U.S. Representatives and Senators to make them aware of the impact of FM on their lives and to urge them to direct more money towards research on it.

Thanks in part to lobbyist Bruce Cameron of Washington, D.C., who has worked for the interests of those with FM, both houses adopted specific language on FM and urged NIAMS to "consider additional appropriate steps." Including, the requests for applications to strengthen the NIH research effort in this disease area. The House and Senate Appropriations Committees "encourages NIAMS, specifically, and NIH, generally, to give additional appropriate steps, including an RFA (Request for Applications) on fibromyalgia, to increase the research that is carried out on this widespread and debilitating disease."

Research funding for 1998 is $13,600,000 for CFS; $20,080,000 for rheumatoid arthritis; and only $2,350,000 for FM (*Fibromyalgia Times*, Summer 1997). The per person costs given previously are from these figures. Each year the two appropriations committees meet during the spring and summer to decide on funding budgets, so each year you will have an opportunity to let your representatives know how you feel about fibromyalgia and the need for more funds.

(One interesting note to add. As I was running through some of the web sites the night before I finished this manuscript, I found a notice that Dr. Katz, in addressing Congress on the 1999 budget, mentioned the Mid-Atlantic Conference on Fibromyalgia Treatment and made a case for increased funding.)

So how can you help bring about a political or social change with regard to fibromyalgia? Your first step should be to actively support one of the national associations that work to provide education and support for fibromyalgia. The four oldest and most active are: Fibromyalgia Alliance of America, Inc., (FMAA), P.O. Box 21990, Columbus, OH 43221–0990, Phone: (614) 457–4222, $25 annual membership fee which includes the quarterly newsletter. National Fibromyalgia Research Association (NFRA), P.O. Box 500, Salem, OR 97308, *http://www.teleport.com/~nfra* or *nfra@teleport.com*. NFRA is "dedicated to education, treatment and finding a cure for fibromyalgia" and is primarily an FMS activist group. The Fibromyalgia Association of Greater Washington, Inc. (FMAGW), 13203 Valley Drive, Woodbridge, VA 22191–1531, Phone: (703) 790–2324, Fax: (703) 273–9208, FMAGW also publishes a quarterly newsletter. Membership is $25 for the first year with renewal rates now at $23/yr; members receive a discount on publications and tapes available through FMAGW and on registration fees for programs.

Kristin Thorson began publishing *The Fibromyalgia Network*, a quarterly newsletter in April, 1988 and also provides several publications of interest to those with FM. The newsletter's address is: P.O. Box 31750, Tucson, AZ 85751–1750, Toll-free: (800) 853–2929, Fax: (520) 290–5555, *www.fmnetnews.com*. The American Fibromyalgia Syndrome Association was organized by Kristen to provide funding for projects on fibromyalgia, chronic fatigue syndrome, and related disorders: 6380 E. Tanque Verde Road, Ste D, Tucson, AZ 85715, Phone (520) 733–1570, Fax: (520) 290–5550.

Each of these organizations (and many more than I haven't listed) work very hard to increase awareness of fibromyalgia and raise funds for research. Essentially they do much of the same things, with some minor differences between the associations: publish newsletters (with the exception of the NFRA), provide information packets for individuals as well as doctors and legal representatives, hold or help sponsor programs and conferences on fibromyalgia and related topics, as well as seek donations to aid in funding research. NFRA has a fibromyalgia awareness pin which they sell for $5 with all proceeds going to research. Members of the organizations meet with leading doctors and researchers as well as members of the government to advocate on the behalf of those with fibromyalgia. While these organizations have made giant steps forward for us, they need your help as well. It costs money to produce, print, and mail newsletters, and brochures and information packets.

What can you do? First, become a member of one or all of the groups. This will keep you up-to-date on the latest information on FM by way of their newsletters or regular mail-outs. Second, support your own local group, if you have one. The more people who are informed about fibromyalgia, the better chance we will have to convince those who control funding that we make up a significant portion of the vot-

ing public. At the conservative figure of 2 percent of the population, or 5 to 6 million, we have the numbers to have an impact.

The third step is to make a donation to one of the organizations listed above or to one of the researchers indicated in the previous section. If you can afford it, you might try to budget a yearly amount. Remember, when you make a donation to a 501(c)(3), non-profit organization, that donation is tax-exempt and can be claimed as a deduction on your income tax.

Fourth, you might consider organizing some sort of fund-raiser with the proceeds going to research. There are a number of ways to raise funds and they can be on a local, regional, or national level. You might find ideas of what others are doing by reading the newsletters.

Lastly, and probably most importantly, you should contact your Congressional Representative and Senators to urge them to support funding for research. Kristin Thorson, in *Fibromyalgia Network* (January 1998), recommended that readers contact their Congressmen by mail, telephone or Internet. You can find the names and addresses in your phone book under U.S. Government, in your local newspaper, or you can call the Capital Hill switchboard at (202) 224–3121 for their Washington phone number. The generic address for all U.S. Representatives is:

The Honorable (name of Representative)
U.S. House of Representatives
Washington, D.C. 20515

Or for senators:
The Honorable (name of Senator)
U.S. Senate
Washington, D.C. 20510

You also might write to the following National Institutes of Health Directors, urging them to increase funding:

Audrey S. Penn, M.D.
Acting Director, NINDS
NIH Bldg 31, Room 8A52
31 Center Drive M.S.C. 2540
Bethesda, MD 20892–2540
Phone: (301) 496–3167
Fax: (301) 496–0296
ap101d@nih.gov

Stephen Katz, M.D. Ph.D.
Director, NIAMS
NIH Bldg 31, Room 4C32
31 Center Drive M.S.C. 2350
Bethesda, M.D. 20892–2350
Phone: (301) 496–4353
Fax: (301) 480–6069

It may seem that one person alone can accomplish very little and yet when one person donates even a little, such as $1, that action, combined with the actions of many other individuals can add up to quite a bit. If every one of the estimated 10 million Americans who have FM donated $1 for research, that would be $10 million. Because most of those 10 million individuals probably have at least one family member or friend who cares enough about them to also donate $1, you would have $20 million. In fact, the actions of people like Kristin, who has FM, and Jack Scott, who is the man who is responsible for the National Fibromyalgia Research Association (NFRA) because his wife has fibromyalgia, have raised far more than $1 each. The American Fibromyalgia Syndrome Association's goal for 1998 is $150,000 for research. The NFRA has been a major factor in obtaining funds for FM research, and has acted as sponsors and co-sponsors of patient and physician conferences.

One person can accomplish something, especially when he or she works with others to achieve the same goal. I urge you to take as many of the steps I've listed above as you can, financially and personally. In fact, I challenge you to do something today to help increase awareness of fibromyalgia, assist in covering the costs of providing information to individuals who can't afford a membership or a subscription, or contact your Senator or Representative or the NIH and NIAMS directors about fibromyalgia. In the end, we are all winners.

There are many more people than I have mentioned here who have given a great deal of themselves, their time, and their money on behalf of FM. I respect all of those individuals, especially those with fibromyalgia who give of themselves, because it is hard enough to take care of our daily activities, much less give even more time and energy. I would like to say thank you to all of those who have worked hard in the associations listed above, in regional or local support groups, or on an individual basis. As Mary Anne Saathoff at the Fibromyalgia Alliance of America said everyone can contribute their efforts and have their role. Because the need is very great, it takes all of us, doing what we each do best, to meet the goals of education and activism and eventually, to find the cause and an effective treatment for FM.

STATE VOCATIONAL REHABILITATION OFFICES

Call the following state offices for the office you should go to in order to apply for benefits.

Alabama
Division of Rehabilitation Services
P.O. Box 11586
2129 East South Boulevard
Montgomery, Alabama
36111–0586
Phone: (334) 281–8780

Alaska
Division of Vocational Rehabilitation
801 West 10th Street, Suite 200
Juneau, Alaska 99801–1894
Phone: (907) 465–2814

Arizona
Rehabilitation Services Administration
Department of Economic Security
1789 West Jefferson, 2nd Floor, NW
    Wing
Phoenix, Arizona 85007
Phone: (602) 542–3332

Arkansas
Division of Rehabilitation Services
Department of Human Services
1616 Brookwood Drive
Little Rock, Arkansas 72202
Phone: (501) 296–1600

California
Department of Rehabilitation
830 K Street Mall, Room 307
Sacramento, California 95814
Phone: (916) 445–3971

Colorado
Rehabilitation Services
1575 Sherman Street, 4th Floor
Denver, Colorado 80203
Phone: (303) 866–5196

Connecticut
Department of Human Resources
Bureau of Rehabilitation Services
25 Sigourney Street, 11th Floor
Hartford, Connecticut 06106
Phone: (860) 723–1400

Delaware
Division of Vocational Rehabilitation
4425 N. Market Street
Fox Valley P.O. Box 9969
Wilmington, Delaware 19809–0969
Phone: (302) 761–8300

District of Columbia
D. C. Rehabilitation Services
Commission on Social Services
Department of Human Services
800 9th Street, S.W., 4th Floor
Washington, D.C. 20024
Phone: (202) 645–5703

Florida
Division of Vocational Rehabilitation
Department of Labor and Employment
    Security, Building A
2002 Old St. Augustine Rd.
Tallahassee, Florida 32399–0696
Phone: (904) 488–6210

Georgia
Division of Rehabilitation
Department of Human Services
10 Park Place South, Suite 602
Atlanta, Georgia 30303
Phone: (404) 657–4726

Hawaii
Division of Vocational Rehabilitation
Department of Human Services
P.O. Box 339
1000 Bishop Street
Suite 605
Honolulu, Hawaii 96809
Phone: (808) 586–5355

Idaho
Division of Vocational Rehabilitation
Len B. Johnson Building, Room 150
P.O. Box 83720
Boise, Idaho 83720–0096
Phone: (208) 334–3390

Illinois
Department of Rehabilitation Services
623 East Adams Street
P.O. Box 19429
Springfield, Illinois 62794–9429
Phone: (217) 782–2093

Indiana
Department of Human Services
P.O. Box 7083
402 West Washington Street Suite
    W451
Indianapolis, Indiana 46204–7083
Phone: (317) 232–7000

Iowa
Div. of Vocational Rehabilitation
    Services
Department of Education
510 East 12th Street
Des Moines, Iowa 50319
Phone: (515) 281–4311

Kansas
Division of Rehabilitation Services
3640 S W Topeka Boulevard, Suite 150
Topeka, Kansas 66611–2373
Phone: (785) 267–5301

Kentucky
Department of Vocational
    Rehabilitation
Bureau of Rehabilitation Services
209 St. Clair Street
Frankfort, Kentucky 40601
Phone: (502) 564–4566

Louisiana
Division of Rehabilitation Services
Office of Community Services
8225 Florida
Baton Rouge, Louisiana 70806
Phone: (504) 925–4131

Maine
Bureau of Rehabilitation
Department of Human Services
35 Anthony Avenue
Augusta, Maine 04333–0150
Phone: (207) 624–5300

Maryland
Division of Vocational Rehabilitation
State Department of Education
2301 Argonne Drive
Baltimore, Maryland 21218
Phone: (410) 554–9385
Toll-free: (888) 554–0334

Massachusetts
Rehabilitation Commission
27–43 Wormwood Street 6th Floor
Boston, Massachusetts 02210
Phone: (617) 727–2183

Michigan
Michigan Rehabilitation Services
P.O. Box 30010
608 West Allegan Street
Lansing, Michigan 48909
Phone: (517) 335–1343

Minnesota
Division of Rehabilitation Services
390 North Robert Street 1st Floor
St. Paul, Minnesota 55101
Phone: (612) 296–5616

Mississippi
Vocational Rehabilitation Services
1281 Highway 51
Madison, Mississippi 39110
Phone: (601) 853–5100

Missouri
State Department of Education
Division of Vocational Rehabilitation
3024 W. Truman Boulevard
Jefferson City, Missouri 65109
Phone: (573) 751–3251

Montana
Dept. of Social and Rehabilitation
    Services
Rehabilitation Services Division
P.O. Box 4210, 111 Sanders
Helena, Montana 59604
Phone: (406) 444–2590

Nebraska
Division of Rehabilitation Service
State Department of Education
301 Centennial Mall, South 6th Floor
P. O. Box 94987
Lincoln, Nebraska 68509
Phone: (402) 471–3652

Nevada
Department of Human Resources
    Rehabilitation Division
505 East King Street, Room 502
Carson City, Nevada 89710
Phone: (702) 687–4440

New Hampshire
Division of Vocational Rehabilitation
State Department of Education
78 Regional Drive, Bldg. 2
Concord, New Hampshire 03301
Phone: (603) 271–3471

New Jersey
Div. of Vocational Rehabilitation
    Services
Department of Labor and Industry
135 East State St., P.O. Box 398
Trenton, New Jersey 08625–0398
Phone: (609) 292–5987

New Mexico
Division of Vocational Rehabilitation
State Department of Education
435 Saint Michaels Drive Building D
Santa Fe, New Mexico 87505
Phone: (505) 954–8511

New York
Office of Vocational and Educational
    Services for Individuals with
    Disabilities
Room 1606 1 Commerce Plaza
99 Washington Avenue
Albany, New York 12234
Phone: (518) 474–2714

North Carolina
Div. of Vocational Rehabilitation
    Services
Department of Human Resources
P.O. Box 26053
805 Ruggles Drive
Raleigh, North Carolina 27611
Phone: (919) 733–3364

North Dakota
Office of Vocational Rehabilitation
Department of Human Services
Dacotah Building 600
South 2nd Street, Suite 1B
Bismark, North Dakota 58504–5729
Phone: (701) 328–8950

Ohio
Ohio Rehabilitation Services
    Commission
400 East Campus View Boulevard
Columbus, Ohio 43235–4604
Phone: (614) 438–1210

Oklahoma
Division of Rehabilitation
Department of Human Resources
3535 NW 58th Street, Suite 500
Oklahoma City, Oklahoma 73112
Phone: (405) 951–3400

Oregon
Division of Vocational Rehabilitation
Department of Human Services
500 Summer Street, NE
Salem, Oregon 97310–1018
Phone: (503) 945–5880

Pennsylvania
Office of Vocational Rehabilitation
1300 Labor and Industry Building
7th and Forster Streets
Harrisburg, Pennsylvania 17120
Phone: (717) 787–5244

Rhode Island
Vocational Rehabilitation
Department of Human Services
40 Fountain Street
Providence, Rhode Island 02903
Phone: (401) 421–7005

South Carolina
Vocational Rehabilitation Department
1330 Boston Avenue
West Columbia, South Carolina 29171
Phone: (803) 896–6333

South Dakota
Division of Vocational Rehabilitation
East Highway 34
c/o 500 East Capitol
Pierre, South Dakota 57501–5070
Phone: (605) 773–3195

Tennessee
Division of Rehabilitation Services
Department of Human Services
Citizens Plaza Building Suite 1100
Nashville, Tennessee 37248–6000
Phone: (615) 313–4891

Texas
Rehabilitation Commission
4900 North Lamar
Austin, Texas 78751–2399
Phone: (512) 424–4420
Toll-free: (800) 628–5115

Utah
Office of Rehabilitation
250 East 5th South
Salt Lake City, Utah 84111
Phone: (801) 538–7530

Vermont
Vocational Rehabilitation Division
Agency of Human Services
Osgood Building, Waterbury Complex
103 South Main Street
Waterbury, Vermont 05671–2303
Phone: (802) 241–2640

Virginia
Department of Rehabilitative Services-
    Commonwealth of Virginia
8004 Franklin Farms Drive
P.O. Box K300
Richmond, Virginia 23288
Phone: (804) 662–7000

Washington
Division of Vocational Rehabilitation
Department of Social and Health
    Services
P.O. Box 45340
Olympia, Washington 98504
Phone: (360) 438–800z0
Toll-free: (800) 637–5627(Washington
    only)

West Virginia
Division of Vocational Services
State Capitol Building, P.O. Box 50890
Charleston, West Virginia 25305–0890
Phone: (304) 766–4601

Wisconsin
Division of Vocational Rehabilitation
Department of Health and Social
    Services
1 West Wilson, 8th Floor
P.O. Box 7852
Madison, Wisconsin 53707
Phone: (608) 243–5600

Wyoming
Division of Vocational Rehabilitation
Department of Health and
    Employment
1100 Herschler Building
Cheyenne, Wyoming 82002
Phone: (307) 777–7385